Fluid Mech[a]
Heat Transfer

Fluid Mechanics and Heat Transfer

Inexpensive Demonstrations and Laboratory Exercises

Edited by
William Roy Penney
Edgar C. Clausen

CRC Press
Taylor & Francis Group
Boca Raton London New York

CRC Press is an imprint of the
Taylor & Francis Group, an **informa** business

CRC Press
Taylor & Francis Group
6000 Broken Sound Parkway NW, Suite 300
Boca Raton, FL 33487-2742

© 2018 by Taylor & Francis Group, LLC
CRC Press is an imprint of Taylor & Francis Group, an Informa business

No claim to original U.S. Government works

Printed on acid-free paper

International Standard Book Number-13: 978-0-8153-7431-2 (Paperback)
International Standard Book Number-13: 978-0-8153-7436-7 (Hardback)

Visit the Taylor & Francis Web site at
http://www.taylorandfrancis.com

and the CRC Press Web site at
http://www.crcpress.com

Contents

Section II   Heat Transfer Experiments and Demonstrations

Preface

One of the major objectives of engineering education is the effective transfer of subject information to the engineering students. Students are likely to encounter a vast array of teaching techniques and styles during their academic careers, and this variety generally adds to the educational experience. However, if students were given a choice, they would most likely select an interactive teaching style for their classes. In a survey of 47 University of Michigan undergraduate engineering students (30 males, 17 females), Pomales-Garcia and Liu (2007) found that the students most preferred teaching that included examples, demonstrations, stories, websites, visual displays, group work, competitions, and oral presentations. As engineering class sizes increase, the temptation for instructors is to move toward lecture-based learning (Hora et al. 2012)—"How can I possibly interact with all of these students?" Notable exceptions to this practice exist, including the flipped classroom.

Student engagement through examples and classroom demonstrations is prominent among the student preferences. Johri and Olds (2011) note that some of the most essential skills in engineering come about through the use of tools and materials, and through interactions with other people. Clearly, engineering students benefit from the *hands-on* aspects of their education such as co-ops/internships, research, and undergraduate laboratories. Despite this fact, there has been a pedagogical shift toward classroom and lecture-based engineering education, and away from laboratory education over the past 30 years (Abdulwahed and Nagy 2009; Hofstein and Lunetta 1982, 2004). Feisel and Rosa (2005) attribute this shift to the increasing complexity and cost of laboratory equipment and the change in motivation of the faculty, but also note that that the integration of the computer into the laboratory has been a positive influence.

Fortunately, the trend away from laboratory instruction is reversing, and laboratory pedagogy is now recognized as a fertile ground for engineering research (Abdulwahed and Nagy 2009). Lin and Tsai (2009) note that learning environments that are student-centered, peer-interactive, and teacher-facilitated significantly help students in learning complex engineering concepts. The literature describes a number of techniques for engaging students in the classroom including the use of virtual engineering laboratories (Mosterman et al. 1994); developing interactive, activity-driven classroom environments (Bidanda and Billo 1995; Carr et al. 1995; Kresta 1998; Lin and Tsai 2009);

relating the curriculum to real-life problems (Finelli et al. 2001); and even using games as teaching tools (Bodnar et al. 2016).

Heat transfer and fluid mechanics have been popular subjects for classroom engagement, both in the laboratory and as classroom demonstrations. Fraser et al. (2007) described the use of computer simulations to enhance both the classroom and laboratory experiences. Wicker and Quintana (2000) extended the use of fluid mechanics to the design and fabrication of lab experiments by the students, and Walters and Walters (2010) used the combination classroom instruction and lab experience to introduce fluid mechanics to talented high school students. Loinger and Hermanson (2002) used an integrated experimental-analytical-numerical approach in the teaching of fluid mechanics, and student surveys showed that 90% of their students preferred this re-designed class to the traditional lecture class, and also felt like they obtained a better understanding of the engineering fundamentals.

Several papers have specifically addressed methods for improving or supplementing the teaching of engineering, including the use of spreadsheets to solve two-dimensional heat transfer problems (Besser 2002), the use of a transport approach in teaching turbulent thermal convection (Churchill 2002), the use of computers to evaluate view factors in thermal radiation (Henda 2004), implementation of a computational method for teaching free convection (Goldstein 2004), and the use of an integrated experimental/analytical/numerical approach that brings the excitement of discovery to the classroom (Olinger and Hermanson 2002). A number of hands-on activities have been suggested for use in the laboratory or classroom, including rather novel experiments such as racecar-based laboratory exercises (Lyons and Young 2001), the drying of a towel (Nollert 2002) and the cooking of french fry-shaped potatoes (Smart 2003).

The purpose of this book is to describe a number of experiments that can be used either in the engineering laboratory or as classroom demonstrations in the teaching of heat transfer or fluid dynamics. These experiments were developed by Professor Roy Penney in his teaching of CHEG 3232, Chemical Engineering Laboratory II, at the University of Arkansas, Fayetteville, Arkansas. The equipment for the vast majority of these experiments can be constructed from materials present in most engineering departments or easily purchased from box stores. Full descriptions of the laboratory equipment and supplies for each experiment are presented, as well as the required experimental protocol and necessary safety issues. The background theory and methods for the analysis and treatment of the experimental data are presented, along with student-generated data and data analyses. Most of the experiments were designed to give results that can be predicted by a mathematical model developed from first principles.

How Do I Use This Book?

All of the experiments in this book can be used as exercises in a laboratory course demonstrating heat transfer and fluid flow. These experiments can be very helpful to supplement the standard repertoire of laboratory experiments, which become well known to students as they are assigned semester after semester. Many of the experiments are very suitable for students to design and build the experimental apparatus; use them to give students hand-on experience in building something. Many of the experiments can be adapted to serve as classroom demonstrations, although required space and the need for water may, in some cases, limit their use. Finally, each experiment has experimental data, generated by students, that might be used in data analysis exercises.

A number of professors have used these experiments in their labs and classrooms over the past 10 years. Dr. Penney (rpenney@uark.edu) and Dr. Clausen (eclause@uark.edu) are available to work with instructors to implement these experiments and to answer any questions regarding experimental equipment, procedures, and results. The computer programs used in these experiments are available upon request from the authors.

References

Abdulwahed, M. and Z.K. Nagy, Applying Kolb's experiential learning cycle for laboratory education, *Journal of Engineering Education*, 98, 3, 283–294, 2009.

Besser, R.S., Spreadsheet solutions to two-dimensional heat transfer problems, *Chemical Engineering Education*, 36, 2, 160–165, 2002.

Bidanda, B. and R.E. Billo, On the use of students for developing engineering laboratories, *Journal of Engineering Education*, 84, 2, 205–213, 1995.

Bodnar, C.A., D. Anastasio, J.A. Enszer, and D.D. Burkey, Engineers at play: Games as teaching tools for undergraduate engineering students, *Journal of Engineering Education*, 105, 1, 147–200, 2016.

Carr, R., D.H. Thomas, T.S. Venkataraman, A.L. Smith, M.A. Gealt, R. Quinn, and M. Tanyel, Mathematical and scientific foundations for an integrative engineering curriculum, *Journal of Engineering Education*, 84, 2, 137–150, 1995.

Churchill, S.W., A new approach to teaching turbulent thermal convection, *Chemical Engineering Education*, 36, 4, 264–270, 2002.

Feisel, L.D. and A.J. Rosa, The role of the laboratory in undergraduate engineering education, *Journal of Engineering Education*, 94, 1, 121–130, 2005.

Finelli, C.J., A. Klinger, and D.D. Budny, Strategies for improving the classroom experience, *Journal of Engineering Education*, 90, 4, 491–497, 2001.

Fraser, D.M., R. Pillay, L. Tjatindi, and J.M. Case, Enhancing the learning of fluid mechanics using computer simulations, *Journal of Engineering Education*, 96, 4, 381–388, 2007.

Goldstein, A.S., A computational model for teaching free convection, *Chemical Engineering Education*, 38, 4, 272–278, 2004.

Henda, R., Computer evaluation of exchange factors in thermal radiation, *Chemical Engineering Education*, 38, 2, 126–131, 2004.

Hofstein, A. and V.N. Lunetta, The role of laboratory in science education: Neglected aspects of research, *Review of Educational Research*, 52, 2, 201–217, 1982.

Hofstein, A. and V.N. Lunetta, The laboratory in science education: Foundations for the twenty-first century, *Science Education*, 88, 1, 28–54, 2004.

Hora, M.T., J. Ferrare, and A. Oleson, Findings from classroom observations of 58 math and science faculty, Wisconsin Center for Education Research, University of Wisconsin-Madison, Madison, MI, 2012.

Johri, A. and B. Olds, Situated engineering learning: Bridging engineering education research and the learning sciences, *Journal of Engineering Education*, 100, 1, 151–185, 2011.

Kresta, S., Hands-on demonstrations: An alternative to full scale lab experiments, *Journal of Engineering Education*, 87, 1, 7–9, 1998.

Lin, C.C. and C.C. Tsai, The relationship between students' conception of learning engineering and their preference for classroom and laboratory learning environments, *Journal of Engineering Education*, 98, 2, 193–204, 2009.

Loinger, D.J. and J.C. Hermanson, Integrated thermal-fluid experiments in WPI's discovery classroom, *Journal of Engineering Education*, 91, 2, 239–243, 2002.

Lyons, J. and E.F. Young, Developing a systems approach to engineering problem solving and design of experiments in a racecar-based laboratory course, *Journal of Engineering Education*, 90, 1, 109–112, 2001.

Mosterman, P.J., M.A.M. Dorlandt, J.O. Campbell, C. Burow, R. Burow, A.J. Broderson, and J.R. Bourne, Virtual engineering laboratories: Design and experiments, *Journal of Engineering Education*, 83, 3, 279–285, 1994.

Nollert, M.U., An easy heat and mass transfer experiment for transport phenomena, *Chemical Engineering Education*, 36, 1, 56–59, 2002.

Olinger, D.J. and J.C. Hermanson, Integrated thermal-fluid experiments in WPI's discovery classroom, *Journal of Engineering Education*, 91, 2, 239–243, 2002.

Pomales-García, C. and Y. Liu, Excellence in engineering education: Views of undergraduate engineering students, *Journal of Engineering Education*, 96, 3, 253–262, 2007.

Smart, J.L., Optimum cooking of french fry-shaped potatoes: A classroom study of heat and mass transfer, *Chemical Engineering Education*, 37, 2, 142–147, 153, 2003.

Walters, K. and K. Walters, Introducing talented high school students via a fluid mechanics short course, *Proceedings of the 2010 American Society for Engineering Education Annual Conference & Exposition*, 2010.

Wicker, R.B. and R. Quintana, An innovation-based fluid mechanics design and fabrication laboratory, *Journal of Engineering Education*, 89, 3, 361–367, 2000.

Editors

William Roy Penney currently serves as Professor Emeritus of Chemical Engineering at the University of Arkansas, Fayetteville, Arkansas. His research interests include fluid mixing and process design, and he has been instrumental in introducing hands-on concepts to the undergraduate classroom. Professor Penney is a registered professional engineer in the state of Arkansas.

Edgar C. Clausen currently serves as Professor and Associate Department Head in Chemical Engineering at the University of Arkansas, Fayetteville, Arkansas. His research interests include bioprocess engineering, the production of energy and chemicals from biomass and waste, and enhancement of the K-12 educational experience. Professor Clausen is a registered professional engineer in the state of Arkansas.

Section I

Fluid Mechanics Experiments and Demonstrations

1

Determining the Net Positive Suction Head of a Magnetic Drive Pump[*]

Allen A. Busick, Melissa L. Cooley, Alexander M. Lopez,
Aaron J. Steuart, William Roy Penney, and Edgar C. Clausen

CONTENTS

1.1 Introduction

In the operation of pumps, cavitation will occur if the difference between the suction pressure and the vapor pressure of the pumped fluid drops below a critical value. Furthermore, "if the suction pressure is less than the vapor pressure, vaporization will occur in the suction line and liquid can no longer be drawn into the pump" (McCabe et al. 2005). During cavitation, some of the liquid flashes to form vapor inside the lowest pressure regions of the pump. These vapor bubbles are then carried to an area of higher pressure, where they suddenly collapse, resulting in the noise associated with cavitation. "Cavitation in a pump should be avoided since it is often accompanied by metal removal, vibration, reduced flow, noise and efficiency loss" (Green and Perry 2008).

---

[*] Reprinted from *Proceedings of the 2010 American Society for Engineering Education Midwest Section Annual Conference*, A.A. Busick, M.L. Cooley, A.M. Lopez, A.J. Steuart, W.R. Penney, and E.C. Clausen, "Determining the Net Positive Suction Head of a Magnetic Drive Pump," Copyright 2010, with permission from ASEE.

Cavitation can be avoided by maintaining or exceeding the required net positive suction head, $NPSH_r$, defined in Equation 1.1 as

$$NPSH_r = \frac{P_a - P^*}{\rho g} - H - H_L \tag{1.1}$$

Since the pressure in the impeller eye will be lower than the pressure in the suction pipe, it is usually necessary to determine $NPSH_r$ experimentally (Munson et al. 2006). Pump manufacturers publish curves relating $NPSH_r$ to capacity and speed for each of their pumps. An example pump curve, including $NPSH_r$, is given on the Crane Engineering website (Schroeder 2014). In applying net positive suction head to fluid system design, the *available NPSH* or $NPSH_a$ must be considered since the $NPSH$ required for operation without cavitation and vibration (or $NPSH_a$) must be greater than the theoretical $NPSH$ (or $NPSH_r$). *Perry's Chemical Engineers' Handbook* (Green and Perry 2008) recommends using the larger of the two $NPSH_a$ values found from Equations 1.2 and 1.3:

$$NPSH_a = NPSH_r + 5\,\text{ft} \tag{1.2}$$

$$NPSH_a = 1.35\,NPSH_r \tag{1.3}$$

The purpose of this paper is to demonstrate a simple and inexpensive laboratory experiment for determining $NPSH_r$ and developed head for a centrifugal pump in a closed circuit. $NPSH_r$ was found by pumping water through the system while also slowly increasing the temperature of the water (to increase the vapor pressure) and observing the onset of cavitation. The lowest temperature causing cavitation was recorded, and the $NPSH_r$ of the pump was determined from Equation 1.1. A secondary, but also very important, purpose of the exercise was to provide students with an opportunity to observe cavitation in a laboratory setting.

1.2 Equipment and Procedures

A flow schematic of the experimental apparatus for measuring $NPSH_r$ is shown in Figure 1.1, and a photograph of the apparatus is shown in Figure 1.2. The pump employed in the experiment was a Magnetek, Model JB15056N, 115 V, 50/60 Hz, 3000 rpm, $\frac{1}{5}$ hp magnetic drive pump. A photograph of the disassembled pump is given in Figure 1.3. The pump was directly connected to the reservoir (a 19 L [5 gal] plastic bucket) with nylon fittings and barbs. This simple coupling arrangement is shown in Figure 1.4. Tygon® tubing (1.9 cm ID × 2.5 cm OD or $\frac{3}{4}$ in ID × 1 in OD) was used to circulate water

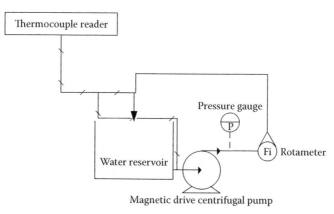

FIGURE 1.1
Schematic flow diagram of the *NPSH* experiment for a magnetic drive centrifugal pump.

FIGURE 1.2
Photograph of the experimental apparatus.

through the system, and 0.6 cm ($\frac{1}{4}$ in) copper tubing was used to inject live steam into the reservoir to raise the temperature of the water. Two wooden 5.1 cm × 10.2 cm (2 in × 4 in) baffles were installed to prevent vortexing into the pump suction line. Type K thermocouples connected to Omega HH82A thermocouple readers were used to monitor the water temperature inside the suction and discharge lines. A 0–15 psig (0–100 kPa above atmospheric) pressure gauge was installed in the discharge line to measure developed

FIGURE 1.3
Photograph of the disassembled pump.

FIGURE 1.4
Photograph of the fittings connecting the pump to the reservoir.

head. Finally, a 1.3 cm ($\frac{1}{2}$ in) ball valve and 0–0.6 $\frac{L}{s}$ (0–10 gpm) rotameter were installed in the discharge line to adjust and monitor the water flow rate.

To initiate the experiment, the pump was started to circulate water through the system, initially with the ball valve in the fully open position. The flow rate and developed pressure head were measured. The water level in the reservoir was maintained at a height of 5 cm (2 in) above the suction centerline of the pump. During system heating, steam was allowed to slowly enter the reservoir to increase the water temperature. During steam addition, the water level in the reservoir tended to increase; thus, occasional water removal was required to maintain the water level at 5 cm. When pump cavitation occurred, the flow rate immediately dropped to zero. No pump noise was heard during cavitation. The obvious conclusion from these observations is that the magnetic coupling was lost when the pump cavitated. When cavitation occurred, the pump and steam were both turned off and the temperature was recorded. After allowing the pump to cool, the pump was once again started. The procedure was repeated at several flow rates by adjusting the ball valve.

1.3 Results and Discussion

Table 1.1 presents the raw experimental data from the cavitation experiments, performed with the Magnetek magnetic drive pump. As expected, the pressure head and temperature at cavitation increased with decreasing flow rate (pump capacity). The variations in cavitation temperature at constant flow rate were most likely due to turbulence at the pump suction or movement of the thermocouple in the suction line during steam addition. These problems could be a focus of future experimentation in order to improve the reproducibility of the experimental results.

TABLE 1.1

Raw Experimental Data

Flow Rate (gpm)	Gauge Reading (psig)	Temperature at Cavitation (°C)
6.8	3.0	89
6.9	3.0	91
6.9	3.0	90
6.9	3.0	91
7.0	3.0	93
5.1	5.8	94
3.1	8.0	97
3.1	8.0	97

The vapor pressure of the water, P^*, may be found from the steam tables at the temperature of cavitation. Alternatively, the appropriate Antoine equation (Himmelblau and Riggs 2004) may be used to calculate the vapor pressure, as given in Equation 1.4:

$$\ln P^* = 18.3036 - \frac{3816.44}{T - 46.13} \tag{1.4}$$

In Equation 1.4, the temperature, T, is in K and the vapor pressure P^*, is in mm Hg. Once the vapor pressure is found, Equation 1.1 may be used to find $NPSH_r$. Assuming the pressure drop in the suction line is small because the length of the line is small with a large diameter, $H_L = 0$. Thus, Equation 1.1 reduces to Equation 1.5:

$$NPSH_r = \frac{P_a - P^*}{\rho g} - H \tag{1.5}$$

The developed head, H_D, is simply the pressure gauge reading (relative to atmospheric pressure, P) converted to head, as noted in Equation 1.6:

$$H_D = \frac{P - P_a}{\rho g} \tag{1.6}$$

where:
$P_a = 735$ mm Hg (14.22 psia or 98 kPa), the observed barometric pressure
$H = 5$ cm (2 in)

Table 1.2 shows $NPSH_r$ and developed head for the Magnetek magnetic drive pump as a function of pump capacity. As is noted, $NPSH_r$ ranged from 0.8 to

TABLE 1.2

$NPSH_r$ and Developed Head as a Function of Pump Capacity

Pump Capacity (gpm)	Temperature at Cavitation (°C)	Water Vapor Pressure (psia)	Water Density $\left(\frac{lb_m}{ft^3}\right)$	$NPSH_r$ (ft)	Developed Head (ft)
6.8	89	9.74	60.25	10.7	7.2
6.9	91	10.50	60.17	8.9	7.2
6.9	90	10.11	60.21	9.8	7.2
6.9	91	10.50	60.17	8.9	7.2
7.0	93	11.32	60.09	6.9	7.2
5.1	94	11.75	60.04	5.9	13.9
3.1	97	13.12	59.91	2.6	19.2
3.1	97	13.12	59.91	2.6	19.2

3.3 m (2.6–10.7 ft), which is reasonable for small pumps, and increased with pump capacity. The developed head ranged from 2.2 to 5.9 m (7.2–19.2 ft), and decreased with pump capacity. As was noted earlier, the variation in cavitation temperature, and thus $NPSH_r$ (average of 2.7 m [9.0 ft], but ranging from 2.1 to 3.3 m [6.9–10.7 ft]) at a capacity of 0.4 $\frac{L}{s}$ (6.9 gpm) was likely due to turbulence at the pump suction or movement of the thermocouple in the suction line during steam addition.

1.4 Conclusions and Recommendations

Within the limits of the equipment, this simple method is fairly accurate and effective in determining the $NPSH_r$ of a pump, and is very effective in demonstrating cavitation and how it occurs in a laboratory setting. It is easier to determine the cavitation point of a magnetic drive pump than a normal centrifugal pump. When the magnetic drive pump cavitates, it stops pumping the liquid completely, while a normal centrifugal pump exhibits a relatively slow decrease in developed head and flow. The experiment can be improved by employing a 3600 rpm pump with a larger impeller, which will cavitate at a lower temperature. With a non-slip drive, a pump curve can be drawn from the data, which will significantly improve the experiment. A manufacturer's pump curve was not available for the small magnetic drive pump used; the use of a pump for which a manufacturer's pump curve is available would add to the learning experience.

1.5 Nomenclature

g Gravitational constant, 9.8 $\frac{m}{s^2}$ $\left(32.2\frac{ft}{s^2}\right)$

H Height of pump above the surface of the fluid in the tank, m (ft)

H_D Developed head, m (ft)

H_L Head loss due to friction in the suction line of the pump, m (ft)

$NPSH_a$ Actual net positive suction head requirement, m (ft)

$NPSH_r$ Required net positive suction head requirement, m (ft)

P_a Pressure at the free liquid surface, barometric pressure, kPa (psi)

P^* Vapor pressure of the fluid at the operating temperature, kPa (psi)

T Temperature, K

P Density of the fluid, $\frac{kg}{m^3}$ $\left(\frac{lb_m}{ft^3}\right)$

References

Green, D.W. and Perry, R.H. 2008. *Perry's Chemical Engineers' Handbook*. 8th ed. pp. 10-27 and 10-28. New York: McGraw-Hill.

Himmelblau, D.M. and Riggs, J.B. 2004. *Basic Principles and Calculations in Chemical Engineering*. 7th ed. Upper Saddle River, NJ: Prentice-Hall.

McCabe, W.L., Smith, J.C. and Harriott, P. 2005. *Unit Operations of Chemical Engineering*. 7th ed. p. 204. New York: McGraw-Hill.

Munson, B.R., Young, D.F. and Okiishi, T.H. 2006. *Fundamentals of Fluid Mechanics*. 5th edn. Hoboken, NJ: John Wiley & Sons.

Schroeder, T. 2014. *How to Read a Centrifugal Pump Curve*. June 26. Accessed August 4, 2017. https://blog.craneengineering.net/how-to-read-a-centrifugal-pump-curve.

2

Bernoulli Balance Experiments Using a Venturi[*]

Megan F. Dunn, William Roy Penney, and Edgar C. Clausen

CONTENTS

[*] Reprinted from Dunn, M.F. et al., *Proceedings of the 2011 American Society for Engineering Education Midwest Section Annual Conference*, Copyright 2011, with permission from ASEE.

2.1 Introduction

Many different types of meters are used industrially to measure the rate at which a fluid is flowing through a pipe or channel. The selection of a meter is based on the applicability of the instrument to the specific problem, its installed costs and operating costs, the desired range of flow rates, and the inherent accuracy of the flow meter (McCabe et al. 2005). In some situations, only a rough indication of the flow rate is required; at other times, highly accurate measurements are necessary to properly control or monitor a process or to facilitate custody transfer.

The venturi meter is one of the more common full-bore meters, that is, meters that operate by sensing all of the fluid in the pipe or channel. In the operation of a venturi meter, upstream and downstream pressure taps are connected to a manometer or differential pressure transmitter. This pressure differential is then used to determine the flow rate by applying a Bernoulli balance between the entry and throat of the meter. The angle of the discharge cone is typically set between 5° and 15° to prevent boundary layer separation and to minimize friction. Typically, 90% of the pressure loss in the upstream cone is recovered, making the venturi very useful for measuring very large flow rates, where power losses can become economically significant. Thus, the higher installed costs of a venturi (over an orifice) are offset by reduced operating costs (McCabe et al. 2005; Green and Perry 2008).

The purpose of this paper is to demonstrate two simple and inexpensive laboratory experiments for analyzing the performance of a venturi. These experiments may be performed as laboratory exercises or as classroom demonstrations. In the first experiment, a shop-fabricated venturi was employed to determine the experimental minor loss coefficient, K, in an unsteady-state system. In the second experiment, a venturi was fashioned from small, inexpensive funnels and once again used to determine experimental minor loss coefficient, this time in a steady-state system.

2.2 Commercial Venturi Experiment

The first experiment employed a shop-fabricated acrylic venturi (Figure 2.1), with a throat diameter of 4.4 mm ($\frac{3}{16}$ in) and a pressure tap in the throat. The venturi was assembled in the flow system of Figure 2.2, consisting of a feed reservoir, an exit reservoir, connecting piping, and plastic tubing that served as simple water-filled manometers. Tap water served as the test fluid.

FIGURE 2.1
Photograph of shop-fabricated venturi.

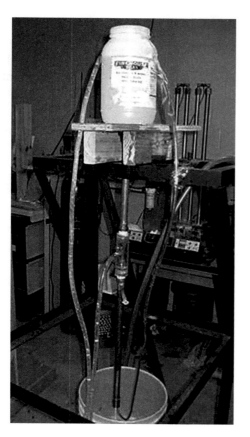

FIGURE 2.2
Flow system with venturi installed.

2.2.1 Experimental Procedures

To ready the system for operation, the top reservoir was filled to its top mark, and the bottom reservoir was completely filled, such that the exit pipe was submerged as shown in Figure 2.3. All air was cleared from the system by suctioning. At time zero, the stopper was removed from the bottom of the piping system, and then the time required for the upper reservoir to drain was recorded as the liquid level passed graduated marks. Several additional measurements were made:

- The distance from the top of the feed reservoir to the water level
- The distance from the top of the feed reservoir to the venturi throat
- The distance from the top of the feed reservoir to the water level in the bottom reservoir
- The diameter of the feed reservoir

2.2.2 Equipment List

- Shop-fabricated venturi, 1 in NPT fittings, equipped with pressure tap at its throat (45° inlet angle, 30° outlet angle, (4 mm [$\frac{3}{16}$ in] throat)
- Water reservoirs (buckets), 2

FIGURE 2.3
System exit in bottom reservoir.

- Copper tubing, 19 mm ($\frac{3}{4}$ in) od, with fittings
- Copper tubing, 6 mm ($\frac{1}{4}$ in) od
- Rubber stopper
- Tygon tubing, 12 mm ($\frac{1}{2}$ in) od
- Stand for apparatus, as available
- Garden hose
- Stopwatch

2.2.3 Results and Discussion

Table 2.1 presents experimental data and Figure 2.4 shows a plot of the height measurements with time as the feed reservoir drained. As is noted in Figure 2.4, the height in the reservoir decreased linearly with time.

TABLE 2.1

Measured Heights

Distance from Feed Reservoir to...	Measured Distance, m
Water level	−1.589
Throat	−0.513
Water level in bottom reservoir	−1.019
Feed reservoir diameter	15.2 cm (6 in)

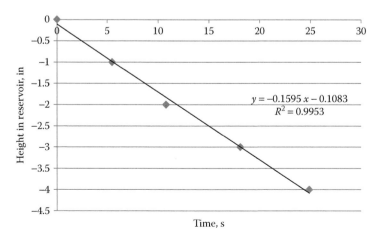

FIGURE 2.4
Height measurements with time as the feed reservoir drained.

2.2.4 Data Reduction

2.2.4.1 Experimental Velocity through the Throat

The velocity through the throat was calculated by dividing the volumetric flow rate from the feed reservoir by the cross-sectional area of the reservoir. The areas of the feed reservoir and venturi throat were determined as in Equations 2.1 and 2.2:

$$A_{fr} = \frac{\pi D_{fr}^2}{4} \tag{2.1}$$

$$A_t = \frac{\pi D_t^2}{4} \tag{2.2}$$

Since the height in the feed reservoir decreased linearly with time, the rate of change may be calculated by Equation 2.3:

$$\frac{dH}{dt} = \frac{\Delta H}{\Delta t} \tag{2.3}$$

The volumetric flow rate from the feed reservoir is then calculated in Equation 2.4 as:

$$Q = \frac{dH}{dt} A_{fr} \tag{2.4}$$

Finally, the throat velocity is obtained by Equation 2.5:

$$v_{vc} = \frac{Q}{A_t} \tag{2.5}$$

2.2.4.2 Throat Velocity from the Bernoulli Balance

McCabe et al. (2005) give the Bernoulli balance (with α terms set to zero) as noted in Equation 2.6:

$$\frac{P_1}{\rho} + gZ_1 + \frac{V_1^2}{2} + W = \frac{P_2}{\rho} + gZ_2 + \frac{V_2^2}{2} + h_f \tag{2.6}$$

In applying the Bernoulli balance across the manometer, the pressure at the throat is obtained in terms of the elevations, with W and h_f both equal to zero, yielding Equation 2.7:

$$\frac{P_m}{\rho} + gH_m = \frac{P_t}{\rho} + gH_t \tag{2.7}$$

But $P_m = 0$, because the reference pressure is atmospheric. Thus, Equation 2.7 yields Equation 2.8, as follows:

$$\frac{P_t}{\rho} = g(H_m - H_t) \qquad (2.8)$$

The Bernoulli balance is now applied from the liquid level in the feed reservoir to the throat, with $W = 0$ and friction neglected between the feed reservoir and the throat, yielding Equation 2.9:

$$\frac{P_{fr}}{\rho} + gH_{fr} = \frac{P_t}{\rho} + gH_t + \frac{V_t^2}{2} \qquad (2.9)$$

Making use of Equation 2.8, with $P_{fr} = 0$, Equation 2.9 becomes Equation 2.10:

$$gH_{fr} = \frac{P_t}{\rho} + gH_t + \frac{V_t^2}{2} = g(H_m - H_t) + gH_t + \frac{V_t^2}{2} \qquad (2.10)$$

The height in the feed reservoir is the reference point, however. Thus, the experimental velocity through the throat is finally obtained in Equation 2.11:

$$v_t = (2gH_m)^{\frac{1}{2}} \qquad (2.11)$$

2.2.4.3 Permanent Friction Loss in the Venturi

In applying the Bernoulli balance between the liquid level in the feed reservoir and the liquid level in the discharge reservoir, the relationship in Equation 2.12 is obtained:

$$h_f = gH_p \qquad (2.12)$$

The permanent friction loss is normally correlated in terms of the velocity in the throat, as is shown in Equation 2.13:

$$h_f = K\frac{V_t^2}{2} \qquad (2.13)$$

where K is the minor loss coefficient for the venturi. Equating Equations 2.12 and 2.13 yields Equation 2.14:

$$K = \frac{2gH_p}{V_t^2} \qquad (2.14)$$

2.2.5 Discussion of Results

2.2.5.1 Throat Velocity

The experimental velocity from Equation 2.5 is 4.82 $\frac{m}{s}$ (15.8 $\frac{ft}{s}$) and the velocity from the application of the Bernoulli balance (Equation 2.11) is 5.58 $\frac{m}{s}$ (18.3 $\frac{ft}{s}$). The experimental velocity should be less than the Bernoulli balance velocity because the velocity near the venturi walls is lower than the centerline velocity. The experimental velocity is lower than the Bernoulli balance velocity by about 16%.

2.2.5.2 Venturi Minor Loss Coefficient

The experimental minor loss coefficient from Equation 2.14 is 0.86, which indicates that only about 15% of the energy content of the vena contracta stream is recovered. This is a low efficiency, which is a result of the large 30° included angle of the diffuser. Duggins (1977) found that the pressure recovery efficiency for a 30° included angle conical diffuser was about 0.25 for an area ratio of 3.265 (a diameter ratio of 1.8). The area ratio for the current diffuser was about $(0.75 \text{ in}/\frac{3}{16} \text{ in})^2 = 16$. With the flow separation that occurs in a 30° included angle diffuser, the outlet diffuser on the shop-fabricated venturi is too short and has too wide an angle for efficient pressure recovery.

2.2.6 Conclusions

1. The throat velocity determined by the Bernoulli balance was about 16% higher than the measured experimental velocity, probably caused by
   a. The boundary layer effects in the small (4 mm, or $\frac{3}{16}$ in) throat
   b. The effect of the jet vena contracta occurring somewhat downstream of the throat of the venturi
2. The minor loss coefficient for the venturi was about 0.86, which indicates only about 14% of the energy contained in the vena contracta stream was recovered. This value is consistent with results reported in the literature.

2.3 Funnel Venturi Experiment

A venturi experiment was also performed with a *homemade* venturi, fashioned from 6.7 cm (2 $\frac{5}{8}$ in) diameter (at the top) plastic funnels. Figure 2.5 shows a photograph of the supplies used to construct the venturi. Epoxy was used to

FIGURE 2.5
Supplies used in preparing a funnel venturi.

securely fasten the top funnel into a 6.7 cm (2 $\frac{5}{8}$ in) hole at the bottom of a bucket. The second funnel was then attached to the first funnel using epoxy with the aid of a 10 cc syringe. Tygon® tubing was split and then attached to the ends of each funnel with putty. Hose clamps were also placed on the ends to further secure the tubing. The split in the tube was repaired by using clear packaging tape. A 9.5 mm (0.375 in) tube was placed through the Tygon® tubing at the throat of the venturi to measure the pressure with a vacuum gauge. A PVC pipe splint and putty were used to support the tube and keep it straight.

2.3.1 Experimental Procedures

The experimental setup for the funnel venturi system was slightly different from the procedures used with the commercial venturi (Figure 2.6), primarily in the use of a vacuum gauge to measure pressure in the throat and the use of a constant-level feed reservoir (operation at steady state). Figure 2.7 shows a close-up of the pressure tap at the throat.

To begin the experiment, the feed and bottom reservoirs (buckets) were filled to the top with tap water. The water flow from the tap was adjusted to keep the height of the water in the feed reservoir constant. Once the system was operating at steady state, the distance from the water level in the feed reservoir to the water level in the bottom reservoir was measured, and the pressure at the throat of the venturi was recorded. The flow rate of water was determined by diverting the feed hose into a reservoir (bucket) for a measured time period. This procedure was repeated at two additional flow rates.

FIGURE 2.6
Flow system with venturi.

FIGURE 2.7
Close-up of venturi tap at the throat.

2.3.2 Equipment List

- Funnels, two, 6.7 cm ($2\frac{5}{8}$ in) top diameter, 10 mm (0.386 in) nozzle diameter
- Water reservoirs (buckets), three 18.9 L (5 gal) pails
- Balance
- Epoxy
- Plumber's putty
- Duct tape
- Packaging tape
- Electrical tape
- Syringe, 10 cc, to dispense epoxy
- $1\frac{1}{2}$ in PVC pipe splint
- Hose clamps
- Tygon® tubing, 35 mm ($1\frac{3}{8}$ in) od
- Copper tubing, 6 mm ($\frac{1}{4}$ in)
- Syphon
- Stand for apparatus, as available
- Vacuum gauge, 0–30 in Hg
- Garden hose
- Stopwatch

2.3.3 Experimental Data

The data from the three experimental runs are presented in Table 2.2. The units are mixed (SI, English units) because they represent the actual experimental data taken in the laboratory. As expected, the vacuum in the vena contracta increased as the flow rate of water increased.

TABLE 2.2

Experimental Data from the Funnel Venturi Experiment

Run	P_t, in Hg (gauge)	H_p, m	Water Collected, lb_m	Time for Collection, s
1	3.7	−0.560	44.5	39.79
2	3.0	−0.495	45.5	49.24
3	1.4	−0.307	43.0	55.16

2.3.4 Data Reduction

2.3.4.1 Experimental Velocity through the Throat

The velocity in the throat is once again found using Equation 2.5. However, in these steady state experiments, the volumetric flow rate is found by Equation 2.15:

$$Q = \frac{m\rho}{t} \tag{2.15}$$

2.3.4.2 Permanent Friction Loss in the Venturi

The minor loss coefficient for the venturi is once again found by Equation 2.14.

2.3.4.3 Bernoulli Balance to Find Throat Pressure

A Bernoulli balance may be written for the system between the top of the feed reservoir and the throat of the venturi, yielding Equation 2.16:

$$\frac{P_{fr}}{\rho} + gH_{fr} = \frac{P_t}{\rho} + gH_t + \frac{V_t^2}{2} \tag{2.16}$$

After rearrangement, the theoretical pressure in the throat can be calculated using Equation 2.17:

$$p_t = \rho \left[\frac{P_{fr}}{\rho} + g\left(H_{fr} - H_t\right) - \frac{V_t^2}{2} \right] \tag{2.17}$$

The experimental absolute pressure in the throat can be found by subtracting the vacuum gauge pressure from atmospheric pressure. Comparisons can then be made between the theoretical and experimental pressures.

2.3.5 Discussion of Results

Table 2.3 shows results for the velocity in the throat vena contracta, the minor loss coefficient, and experimental and theoretical throat pressures for the

TABLE 2.3

Reduced Results from the Experiments

Run	$v_t \frac{m}{s}$	K	P_t, in Hg abs	
			Experimental	Theoretical
1	6.72	0.25	26.22	25.02
2	5.56	0.31	26.92	26.93
3	4.69	0.27	28.52	27.69

three experimental runs. The calculated velocities were 6.72, 5.56, and 4.69 $\frac{m}{s}$ for the three experimental runs. The calculated minor loss coefficients were 0.25, 0.31, and 0.27. Although the coefficient was not constant for the three runs as expected, the percent deviation for these values was only 3.5%, an acceptable experimental error. The experimentally obtained throat pressures for each flow rate were 88.7, 91.1, and 96.6 kPa (26.22, 26.92, and 28.52 in Hg), and the theoretical throat pressures calculated from experimental velocities were 84.7, 92.7, and 93.7 kPa (25.02, 26.93, and 27.69 in Hg). The theoretical and experimental pressures for each data point agreed within 5%, again an acceptable experimental error.

2.3.6 Conclusions

1. The calculated minor loss coefficients were 0.25, 0.31, and 0.27, a scatter of 3.6%.

2. The experimentally obtained throat pressures for each flow rate were 88.7, 91.1, and 96.6 kPa (26.22, 26.92, and 28.52 in Hg), while the theoretical throat pressures based on experimental velocities were 84.7, 92.7, and 93.7 kPa (25.02, 26.93, and 27.69 in Hg). The theoretical and experimental pressures agreed within 5%.

2.4 Nomenclature

A_{fr}	Area of feed reservoir, m²
A_t	Area of throat, m²
D_{fr}	Diameter of feed reservoir, m
D_t	Diameter of throat, m
H	Height, m
H_{fr}	Height of liquid in the feed reservoir, m
H_m	Distance from the top of the feed reservoir to the water, m
H_p	Distance from the top of the feed reservoir to the water in the bottom reservoir, m
H_t	Distance from the top of the feed reservoir to the throat, m
K	Minor loss coefficient for the venturi, dimensionless
P_1	Pressure at position 1, atm
P_2	Pressure at position 2, atm
P_{fr}, P_m	Pressure at manometer entrance, $(P_{fr} = P_m = \text{atmospheric pressure})$
P_t	Pressure in throat, atm
Q	Volumetric flow rate from feed reservoir, $\frac{m^3}{s}$
v_1	Velocity at position 1, $\frac{m}{s}$

v_2	Velocity at position 2, $\frac{m}{s}$
v_t	Velocity in throat, $\frac{m}{s}$
W	Work done on/by the system, $\frac{m^2}{s^2}$
Z_1	Height at position 1, m
Z_2	Height at position 2, m
$\frac{dH}{dt}$	Height change in the feed reservoir with time, $\frac{m}{s}$
g	Acceleration of gravity, $\frac{m}{s^2}$
h_f	Friction in the system, $\frac{m^2}{s^2}$
m	Mass of collected water, kg
t	Time, s
Δ	Change in...
ρ	Fluid (water) density, $\frac{kg}{m^3}$

References

Duggins, R.K. 1977. *Some Techniques for Improving the Performance of Short Conical Diffusers*. December 5–9. Accessed August 4, 2017. http://people.eng.unimelb.edu.au/imarusic/proceedings/6/Duggins.pdf.

Green, D.W. and Perry, R.H. 2008. *Perry's Chemical Engineers' Handbook*. 8th ed. New York: McGraw-Hill.

McCabe, W.L., Smith, J.C. and Harriott, P. 2005. *Unit Operations of Chemical Engineering*. 7th ed. New York: McGraw-Hill.

3

Force Produced by Impingement of a Fluid Jet on a Deflector*

Daniel R. Miskin, William Roy Penney, and Edgar C. Clausen

CONTENTS

3.1 Introduction

Turbines are often used to recover energy from a fluid. In using a Pelton wheel turbine, energy is recovered by directing the fluid onto vanes or buckets. The force due to the impinging fluid rotates the turbine, and this rotation is used to convert the mechanical energy of the flowing fluid into mechanical

---

power that can be used to power an electrical generator (Frederick Institute of Technology undated). The force of the impinging jet of liquid on the vanes or buckets must be calculated to determine the output of the turbine.

Another area of interest to engineers is the force produced by flowing fluids in pipelines. The venting of water from a vessel at 21.8 kPa gauge (150 psig) could produce a water jet velocity of 45 $\frac{m}{s}$ (149 $\frac{ft}{s}$). This velocity through a 10 cm (4 in) vent line would have a flow rate of 1249 $\frac{m^3}{hr}$ (5,500 $\frac{gal}{min}$ or gpm) and produce a resultant force of 21,000 N (4,700 lb$_f$) on a 90° elbow, as calculated by The Engineering Toolbox (undated). A force of this magnitude would require careful support of the piping to prevent failure from the bending moment on the pipe.

The specific objectives of this activity were to:

- Experimentally determine the flow rate at which an impinging fluid jet would lift a deflector
- Develop a simple mathematical model to predict the force of the jet
- Compare the predicted force from the mathematical model to the force obtained from the experimental data

3.2 Equipment and Procedures

The following paragraphs describe the equipment and materials, experimental procedures, and safety considerations in performing the experiment.

3.2.1 Equipment List

The equipment and supplies used in the experiment were as follows:

- 0.75 kW (1 hp) regenerative pump, from Atrepo USA, serial no. 82009
- Jet deflector, manufactured from a 0.0038 m³ (1 gal) PETE pretzel container, installed with a center pipe
- 1.27 cm ($\frac{1}{2}$ in) copper pipe, threaded at one end, 52.7 cm (20.75 in) long
- 0–690 kPa (0–100 psi) pressure gauge, manufactured by Nosha
- Rotameter, Brooks Instrument Division, Emerson Electric Co., Model: 1305D10A3A1A
- 2.5 cm (1 in) PVC ball valve
- Generic meter stick
- Carpenter's hammer
- 0.019 m³ (5 gal) polyethylene pail

- 1.9 cm ($\frac{3}{4}$ in) galvanized steel pipe nipple, 5.1 cm (2 in) long
- 1.3 cm ($\frac{1}{2}$ in) brass pipe nipple, 7.6 cm (3 in) long, 2
- Aluminum bar for support of copper tube, 2, 15.2 cm × 5.1 cm × 1.6 cm (6 in × 2 in × 0.625 in)
- Stainless steel disk base to support the copper tube in a vertical upright position, 12.7 cm diameter × 1.6 cm thick (5 in diameter × $\frac{5}{8}$ in thick)
- Assorted galvanized steel washers to serve as weights
- Hose clamp
- 1.9 cm ($\frac{3}{4}$ in) brass union
- Bushing, 1.9 cm male to 1.3 cm female ($\frac{3}{4}$ in male to $\frac{1}{2}$ in female)
- 1.3 cm ($\frac{1}{2}$ in) 90° elbow, galvanized steel, 2
- PVC or Tygon tubing
- Stand for bucket and pump
- Power strip
- Dental floss
- Electrical tape
- Digital calipers
- Laboratory scale

3.2.2 Experimental Apparatus

Figure 3.1 shows a photograph of the experimental apparatus, containing the regenerative pump, carpenter's hammer, dental floss, jet deflector, pail, power strip, pressure gauge, rotameter, tubing, stand, ball valve, and yardstick. Figure 3.2 is a photograph of the jet piping, which includes all of the fittings, aluminum bars, weight, pipe, and hose clamp. Figure 3.3 is a photograph of the jet deflector, which was constructed from a 3.8 L (1 gal) pretzel container, installed with a center tube. The 1.6 cm ($\frac{5}{8}$ in) PVC center pipe was 15.2 cm (6 in) long. All threads (3.2 mm-32 or $\frac{1}{8}$ in-32) were used as supports. Figure 3.4 is a photograph of the placement of the washers on the deflector, used to add weight to the container.

3.2.3 Experimental Setup

The following procedure was used to set up the equipment in preparation for experimentation:

- Measure the dimensions of all of the equipment using the meter stick and digital calipers.
- Weigh the jet deflector and each washer using the laboratory scale.

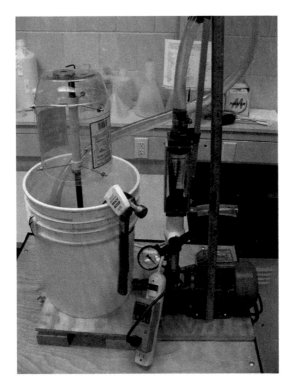

FIGURE 3.1
Photograph of the experimental apparatus.

FIGURE 3.2
Photograph of the jet piping.

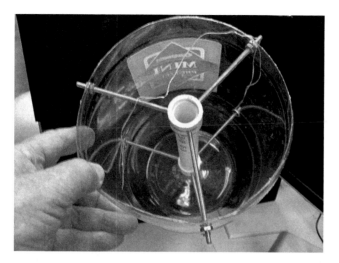

FIGURE 3.3
Photograph of the jet deflector.

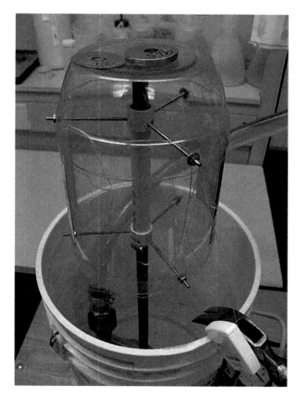

FIGURE 3.4
Photograph of the placement of the washers as weight.

- Pour water into the 3.8 L (5 gal) polyethylene pail until it is half full.
- Place the jet piping into the pail.
- Place the jet deflector on the pipe.
- Tie the jet deflector to the pail using dental floss, but leave the floss loose to allow the jet deflector to float.
- Plug the regenerative pump into the power strip and the power strip into a power source.
- Recruit four people to perform the experiment.
  - Stand one person by the pump to turn it off in case of an upset.
  - Assign one person to adjust the ball valve and place washers on the jet deflector.
  - Assign one person to monitor the jet deflector to determine when it begins to float and take the rotameter readings.
  - Assign one person to record the weight of the jet deflector and washers and the rotameter readings.

3.2.4 Experimental Procedure

The following procedure was used in conducting the experiment:

- Turn on the regenerative pump.
- Using the carpenter's hammer, adjust the ball valve until the jet deflector floats.
- Record the total weight of the jet deflector and the rotameter reading. In performing the experiment, the students took rotameter readings at the top of the float.
- Add two 1.3 cm ($\frac{1}{2}$ in) washers to the top of the jet deflector and repeat the experiment. Do this several times for a total of six experiments.

3.2.5 Safety

As in all laboratory exercises, always wear safety goggles, closed-toe shoes, and pants throughout the experiment.

3.3 Experimental Results

Table 3.1 shows the data from the experiment, shown as the required flow rate for a given total deflector weight. As more weight is added to the deflector, the flow rate of water increases to float the deflector.

TABLE 3.1

Experimental Data

Total Deflector Weight, g	Flow Rate, $\dfrac{m^3}{hr}$	Flow Rate, gpm
204.8	2.2	9.8
239.0	2.3	10.2
273.2	2.5	10.9
307.5	2.6	11.4
341.9	2.7	11.9
367.5	2.8	12.2

3.4 Data Reduction

The force of an impinging jet, which deflects 90°, is given by Equation 3.1 (Çengel and Cimbala 2006):

$$F_j = \rho Q v \tag{3.1}$$

The force due to gravity is given by Equation 3.2:

$$F_g = mg \tag{3.2}$$

The jet deflector will float when the force of the impinging jet, F_j, equals the force due to gravity, F_g, of the jet deflector plus any added weight.

3.4.1 Statistical Analysis

The deviation between experimental force due to gravity, F_g, and the jet force, F_j, may be calculated by Equation 3.3:

$$\Delta R_m = \frac{\left(F_g - F_j\right)}{F_j} \tag{3.3}$$

The coefficient of variance between the force due to gravity and the jet force is then found from Equation 3.4:

$$C_v = \left[\frac{1}{n}\left(\sum \Delta R_m\right)^{0.5}\right] \tag{3.4}$$

3.4.2 Sample Calculations

Using Run 1 in Table 3.1 (a deflector weight of 204.8 g allowing a flow rate of 2.2 $\frac{m^3}{hr}$ or 9.8 gpm), the calculations of Equations 3.1 through 3.4 were performed. The force of the impinging jet, F_j, was found from Equation 3.1 as 2.40 N, where the velocity, v, is the flow rate divided by the cross-sectional area of the tube. The force due to gravity was found from Equation 3.2 as 2.01 N, where the acceleration of gravity is 9.8 $\frac{m}{s^2}$. The deviation between experimental force due to gravity and the jet force, ΔR_m, is calculated by Equation 3.3 as −0.198. Finally, the coefficient of variance, c_v, is calculated by summing the deviations using Equation 3.4.

3.5 Discussion of Results

Table 3.2 shows the force due to gravity based on the mass of the deflector and added weights and the calculated jet force as a function of the volumetric flow rate, as well as the deviation between these results. The results are also plotted in Figure 3.5. As is noted, the jet force was consistently higher than the gravitational force, most likely because the jet did not deflect at exactly 90° and erroneous flow

TABLE 3.2

Gravitational Force and Jet Force as a Function of Flow Rate

Flow Rate, $\frac{m^3}{hr}$ (gpm)	Force, N		
	Gravitational	Jet	ΔR_m
2.2 (9.8)	2.01	2.40	−0.198
2.3 (10.2)	2.34	2.60	−0.112
2.5 (10.9)	2.68	2.97	−0.111
2.6 (11.4)	3.01	3.25	−0.080
2.7 (11.9)	3.35	3.55	−0.058
2.8 (12.2)	3.60	3.73	−0.035

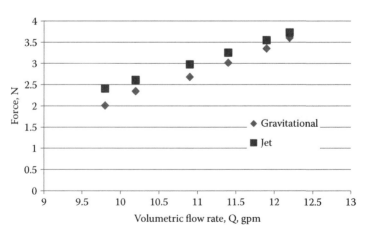

FIGURE 3.5
Plot of gravitational force and jet force as a function of flow rate.

rates were obtained because the rotameter was not calibrated. At the lowest flow rate, there was an error of 19.8% between the two forces. At the highest flow rate, the error was only 3.5%. The coefficient of variance was 5.2%.

3.6 Conclusions

1. The force due to gravity based on the mass of the deflector and added weights and the calculated jet force matched very well (within 3.5%) at high flow rates.

2. The coefficient of variance was 5.2%, which indicates that there are experimental errors that need to be corrected.

3. The errors in the experiment are most likely due to the jet not being deflected at exactly 90° and inaccurate rotameter readings.

3.7 Nomenclature

c_v Coefficient of variance

F_g Force due to gravity, N

F_j Force due to the fluid jet, N

g Acceleration due to gravity, $\frac{m}{s^2}$

m Total mass of the jet deflector and all added weight, kg

n Total number of experimental data points

Q Volumetric flow rate, $\frac{m^3}{s}$

v Velocity of the fluid, $\frac{m}{s}$

ΔR_m Deviation between experimental force due to gravity and the jet force

ρ Density of the fluid, $\frac{kg}{m^3}$

References

Çengel, Y.A. and Cimbala, J.M. 2006. *Fluid Mechanics: Fundamentals and Applications*, p. 238. New York: McGraw-Hill.

The Engineering Toolbox. undated. *Piping Elbows: Thrust Block Forces*. Accessed August 7, 2017. http://www.engineeringtoolbox.com/forces-pipe-bends-d_968.html.

Frederick Institute of Technology. undated. *Impact of a Jet*. Accessed August 7, 2017. http://staff.fit.ac.cy/eng.fm/classes/amee202/Fluids%20Lab%20Impact%20of%20a%20Jet.pdf.

4

Applying the Bernoulli Balance to Determine Flow and Permanent Pressure Loss[*]

Jordan N. Foley, John W. Thompson, Meaghan M. Williams, William Roy Penney, and Edgar C. Clausen

CONTENTS

4.1 Introduction

Isaac Newton said, *"If I have seen further, it is by standing on the shoulders of giants"* (Turnbull 1959). In studying the venturi flowmeter, Daniel Bernoulli (2016) is the first "giant" encountered; he developed his famous Bernoulli equation (efluids 2017) in 1738. The second "giant" is Giovanni Venturi (Wikipedia 2017), who *"was the discoverer of the Venturi effect…in 1797….and… is the eponym for the Venturi tube, the Venturi flow meter and the Venturi pump."* The venturi flowmeter was not applied commercially until Clements Herschel (Herschel 1895; Wikipedia 2016) obtained a U.S. patent for using a *"venturi tube to exercise a suction action"* (Herschel 1888) to measure the flow of

---

[*] Reprinted from Foley, J.N. et al., *Proceedings of the 2015 American Society for Engineering Education Zone III Conference*, Copyright 2015, with permission from ASEE.

water through a pipe. Many publications exist that explain the venturi meter in detail, and there are also several excellent YouTube videos that demonstrate its use and utility (Endress Hauser 2009; Drew 2010).

The major objective of this experiment was to construct, test, and model a simple, inexpensive venturi meter. More specifically, this simple venturi meter, constructed from materials available at a local hardware/auto parts store, was characterized by determining the venturi coefficient and the permanent pressure loss. This experiment is important educationally because it requires students to execute three Bernoulli balances within the overall system—and students often have trouble selecting proper endpoints for Bernoulli balances.

4.2 Experimental

4.2.1 Apparatus

Photographs of the simple venturi apparatus and its components are shown in Figures 4.1 through 4.3. Figure 4.1 presents an overview of the

FIGURE 4.1
An overview of the apparatus.

FIGURE 4.2
A view of the apparatus, as it was operated.

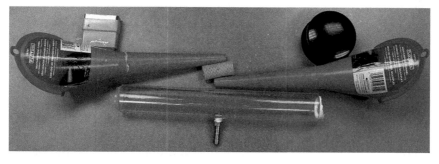

FIGURE 4.3
An exploded view of the venturi, including funnels, connecting tube, and the vacuum housing.

experimental setup and Figure 4.2 presents the apparatus as it was tested. City water was supplied to the apparatus through a garden hose connected to a clothes washer hookup hose. A 1.9 cm ($\frac{3}{4}$ in) pipe tee was placed on the end of the washer hookup hose. A silicone manometer tube was connected to the branch of the tee, and this vertical tube was used to measure the pressure at the end of the hose. A 180 × 1.9 cm ID (70 × $\frac{3}{4}$ in ID) Tygon® tube was used to connect the outlet of the tee to the venturi meter. The ID of the venturi (at the end of the funnels) was measured by inserting a 0.953 cm ($\frac{3}{8}$ in) metal tube, which fit snugly through the ends of the funnels.

An exploded view of the venturi meter is shown in Figure 4.3. Two FloTool funnels (#10701; available at AutoZone) were used to construct the venturi meter. The funnels were connected by a 5.1 cm (2 in)-long section of 1.27 cm ($\frac{1}{2}$ in) ID nylon-reinforced silicone tubing. As the funnels were joined by inserting their ends into the silicone tube, the ends of the funnels were inserted into a 26.7 cm (10.5 in)-long, 3.2 cm ($1\frac{1}{4}$ in) ID Plexiglas® tube; a threaded hole was drilled in the tube center to accommodate a $\frac{1}{8}$ in NPT (~1.0 cm) male-pipe-threaded hose barb. A 0.33 cm ($\frac{1}{8}$ in) hole was drilled through the silicone tubing and through the walls of funnels through the hose barb hole. Silicone caulking was used to seal the joint between the Plexiglas® tube and the funnels. The flowmeter was connected to a wooden stand for stability, as is shown in Figure 4.2. The flowmeter centerline was 48 cm (18 $\frac{7}{8}$ in) above the concrete laboratory floor. The end of the lower funnel of the flowmeter was placed inside a 19 L (5 gal) pail, as is shown in Figures 4.1 and 4.2. A digital vacuum gauge (Dwyer, Model DPG-00, 0–102 kPa (0–30 in Hg)) was used to measure the pressure at the center of the flowmeter.

4.2.2 Experimental Procedure

At the start of an experiment, the end of the 1.9 cm ($\frac{3}{4}$ in) Tygon® tubing was firmly inserted into the mouth of the upper funnel. The valve in the city water line was fully opened at the maximum flow rate to start each of the experimental runs in order to purge the venturi meter of all air. After starting at maximum flow rate, the valve in the city water line was used to adjust the flow rate to the desired levels.

The flow rate was measured experimentally by removing the end of the Tygon® tube from the upper funnel and directing the flow into a tared 4 L flask. The flow was timed using a stopwatch, and the mass of the collected water was measured using an electronic balance. Additional measurements included the height of the water in the inlet line manometer and the vacuum reading at the center of the venturi.

4.3 Experimental Data and Measured Flow Rate

Table 4.1 shows the experimental data and calculated mass flow rates for six runs.

TABLE 4.1

Experimental Data and Calculated Flow Rates

Run	Mass of Water (g)	Measured Time (s)	Calculated Flow Rate $\left(\frac{g}{s}\right)$	Manometer Reading (in water)	Venturi Pressure (mm Hg)	Absolute Venturi Pressure (kPa)[a]
1	4892	7.01	698	91.75	−317	54.7
2	4861	9.03	538	61	−189	72.4
3	4659	13.52	345	40.5	−95	84.3
4	4628	8.19	565	75	−261	62.1
5	4680	7.26	645	89	−310	55.6
6	17064	25.00	683	87	−310	54.0

[a] Barometric pressure was 725 mm Hg.

4.4 Model Development

The modified Bernoulli Balance (commonly called the mechanical energy balance), its development, and its use are explained by Cengel et al. (2012), as is shown in Equation 4.1:

$$\frac{v_1^2}{2g} + H_1 + \frac{P_1}{\rho g} + H_p = \frac{v_2^2}{2g} + H_2 + \frac{P_2}{\rho g} + H_t + H_f \tag{4.1}$$

The velocity, pressure, and elevation terms are applicable at the defined entrance (point 1) and exit (point 2) of the system, and the pump, turbine, and friction terms occur anywhere within the defined system. As was noted earlier, the modified Bernoulli balance must be applied three times to model the venturi flowmeter; thus, the entrances and exits of the three systems will change as the Bernoulli balance is applied to different systems. The inlet and exits of the three systems are defined as

System	Inlet	Point in System	Outlet	Point in System
Inlet line	Tube entrance	1	Tube exit	2
Venturi	Tube exit	2	Venturi	3
Venturi system	Venturi entrance	2	Water level in pail	4

4.4.1 Inlet Line Friction Loss

The Bernoulli balance was applied to the inlet line (i.e., the 1.9 × 180 cm [$\frac{3}{4}$ × 70 in] line) to determine the pressure at the outlet of the inlet tube, and, thus, the pressure at the inlet of the venturi meter. The fluid velocity within the inlet line is determined from the continuity equation, shown in Equation 4.2:

$$v_1 = \frac{M_1}{\rho A_1}$$ (4.2)

where, in Equation 4.3:

$$M_e = \frac{m}{t}$$ (4.3)

and, in Equation 4.4:

$$A_1 = \frac{\pi D_t^2}{4}$$ (4.4)

There is no pump or turbine and $v_1 = v_2$; thus, as is noted in Equation 4.5:

$$P_2 = \rho g \left(H_1 - H_2 + \frac{P_1}{\rho g} - H_f \right)$$ (4.5)

where, as is noted in Equation 4.6:

$$H_f = \frac{f L_t v_1^2}{2 g D_t}$$ (4.6)

The friction factor, f, may be determined from McCabe et al. (2005) for a smooth tube using Equations 4.7 and 4.8:

$$f = 0.0014 + \frac{0.125}{Re^{0.32}}$$ (4.7)

$$Re = \frac{v_1 D_t \rho}{\mu} = \frac{v_t D_t \rho}{\mu}$$ (4.8)

4.4.2 Flow Rate Determination by Venturi Measurements

With no pump or turbine and friction losses ignored in the inlet funnel, the balance reduces to Equation 4.9:

$$\frac{v_2^2}{2g} + H_2 + \frac{P_2}{\rho g} = \frac{v_3^2}{2g} + H_3 + \frac{P_3}{\rho g}$$ (4.9)

Thus, the throat velocity may be calculated as is shown in Equation 4.10:

$$v_3 = \left[2g \left(\frac{v_2^2}{2g} + H_2 + \frac{P_2}{\rho g} \right) - H_3 - \frac{P_3}{\rho g} \right]^{\frac{1}{2}} \qquad (4.10)$$

This calculated throat velocity, v_3, is used in determining the calculated mass flow rate as determined by the venturi flow rate measurement, shown in Equation 4.11:

$$M_3 = M_c = v_3 \rho A_3 \qquad (4.11)$$

where the area of the throat is calculated in Equation 4.12, as

$$A_3 = A_t = \frac{\pi D_3^2}{4} = \frac{\pi D_t^2}{4} \qquad (4.12)$$

The calculated mass flow rate is to be compared with the experimental mass flow rate to determine the discharge coefficient of the venturi meter, where the discharge coefficient is defined in Equation 4.13 as

$$C_d = \frac{M_3}{M_e} = \frac{M_c}{M_e} \qquad (4.13)$$

4.4.3 Correlation for Permanent Pressure Loss by Performing an Overall Bernoulli Balance

Applying the Bernoulli balance to the system defined as that contained between point 2 (the entrance of the venturi meter) and point 4 (the free sur-face of the water in the pail) allows the determination of the friction losses within the venturi meter. Equation 4.1 reduces to Equation 4.14:

$$\frac{v_2^2}{2g} + H_2 + \frac{P_2}{\rho g} = H_4 + H_f \qquad (4.14)$$

Thus, Equation 4.15 results:

$$H_f = \frac{v_2^2}{2g} + H_2 + \frac{P_2}{\rho g} - H_4 \qquad (4.15)$$

Minor losses for fluid fittings and devices are normally correlated by determining the head loss by multiplying a minor loss coefficient by a characteristic velocity head. In this case, the appropriate characteristic velocity head is the velocity head at the venturi throat (i.e., point 4), shown in Equation 4.16:

$$H_f = H_{\text{minor loss}} = K\left(\frac{v_v^2}{2g}\right) = K\left(\frac{v_3^2}{2g}\right) \tag{4.16}$$

Thus, as is noted in Equation 4.17:

$$K = \frac{H_f}{\left(\dfrac{v_v^2}{2g}\right)} = \frac{2gH_f}{v_v^2} \tag{4.17}$$

4.5 Reduced Results and Discussion

The calculated results are summarized in Table 4.2. The experimental mass flow rate is compared to the mass flow rate determined from the Bernoulli balance, showing an average error of 12.5%. The experimental discharge coefficient ranged from 0.82 to 0.92, with the average of 0.89. This discharge coefficient may be compared with discharge coefficients for very well-designed venturi meters that, according to McCabe et al. (2005), "….[are] about 0.98 for pipe diameters of 2–8 in (5.1–20.3 cm) and about 0.99 for larger sizes." The minor loss coefficient varied from 0.27 to 0.33, with an average of 0.28; thus, the permanent pressure loss is about 28% of the velocity head within the throat of the venturi. When compared with the recommendation of McCabe et al. (2005) regarding pressure loss, "typically 90% of the pressure loss in the upstream cone is recovered."

TABLE 4.2

Results for the Mass Flow Measurements and for the Minor Loss Coefficient

Run	Experimental Mass Flow Rate (M_e) $\left(\frac{kg}{s}\right)$	Calculated Mass Flow Rate (M_c) $\left(\frac{kg}{s}\right)$	Mass Flow Error $(M_c$ vs. $M_e)\%$	Orifice Coefficient, C_d	Venturi Minor Loss Coefficient, K
1	0.698	0.762	8.46	0.916	0.283
2	0.538	0.578	6.84	0.931	0.288
3	0.345	0.415	17.0	0.831	0.327
4	0.565	0.687	17.7	0.822	0.279
5	0.645	0.754	14.5	0.855	0.284
6	0.697	0.760	10.6	0.917	0.267
Average			12.5	0.879	0.288

4.6 Conclusions

1. The experiment is an excellent teaching tool because it involves application of three separate Bernoulli balances to reduce the experimental data.

2. The inexpensive venturi meter is not nearly as efficient as a well-designed venturi meter.

3. C_d was in the range of 0.89 and, for a well-designed venturi meter, $C_d \approx 0.98$.

4. The pressure loss is about 30% of the velocity head within the venturi throat; a well-designed meter would experience a 10% loss of the throat velocity head.

5. A more complete experimental program, with several more experimental runs and more duplicate runs, would result in less scatter in the results.

4.7 Nomenclature

Latin Symbols

A_t	Area of the Tygon® tube, m²
A_v	Area of the throat of the venturi, m²
$D_t = D_1$	Diameter of the Tygon® tube, m
$D_v = D_3$	Diameter of the throat of the venturi, m
g	Gravitational constant, $\frac{m}{s^2}$
H_1	Elevation of the laboratory floor, 0 m
H_2	Elevation of the exit of the $\frac{3}{4}$ in Tygon® feed tube above the floor, 0.63 m
$H_3 = H_v$	Elevation of the venturi above the floor, 0.48 m
H_4	Elevation of the water level in the 5 gallon pail above the floor, 0.43 m
H_f	Friction loss in the defined system, m fluid
L_t	Loss in the $\frac{3}{4}$ in Tygon® tube, m
M_c	Calculated mass flow rate of the water through the system, $\frac{kg}{s}$
M_e	Experimental mass flow rate of the water through the system, $\frac{kg}{s}$
P_1	Pressure at Point 1 in the system, that is, the tube entrance, m water
P_2	Pressure at Point 2 in the system, that is, the tube exit, m water

P_3 Pressure at Point 3 in the system, that is, the venturi throat, m H_2O

P_4 Pressure at Point 4 in the system, that is, the water level in the pail, m water

$v_1 = v_v$ Velocity at Point 1 in the system at the tube entrance, $\frac{m}{s}$

$v_2 = v_v$ Velocity at Point 2 in the system at the tube exit, $\frac{m}{s}$

$v_3 = v_v$ Velocity at Point 3 in the system in the venturi throat, $\frac{m}{s}$

v_4 Velocity at Point 4 in the system at the water surface in the pail, $\approx 0 \frac{m}{s}$

v_t Velocity in the Tygon® Tube, $\frac{m}{s}$

Greek symbols

μ Viscosity of water, $\frac{kg}{m\,s}$

ρ Density of water, $\frac{kg}{m^3}$

Dimensionless Parameters

f Friction factor

Re   Reynolds number, $\frac{v_t D_t \rho}{\mu}$

K Minor loss coefficient for friction losses in the venturi meter, $K = \frac{H_f}{\left(v_v^2/2g\right)}$

References

Bernoulli, D. 2016. *Famous Scientists: The Art of Genius.* May 26. Accessed August 7, 2017. http://www.famousscientists.org/daniel-bernoulli/.

Cengel, Y.A., Cimbala, J.M., and Turner, R.H. 2012. *Fundamentals of Thermal Sciences.* chapter 12, Bernoulli and Energy Balances, pp. 112–127. New York: McGraw-Hill.

Drew, B. 2010. *Fluids—Lecture 3.1—Flow Rate Measurement.* November 23. Accessed August 7, 2017. https://www.youtube.com/watch?v=7xUdPVpafyI.

efluids. 2017. *Bicycle Aerodynamics: 6. Bernoulli's Equation.* August 7. Accessed August 7, 2017. http://www.princeton.edu/~asmits/Bicycle_web/Bernoulli.html.

Endress Hauser. 2009. *The Differential Pressure Flow Measuring Principle (Orifice-Nozzle-Venturi).* November 26. Accessed August 7, 2017. https://www.youtube.com/watch?v=oUd4WxjoHKY.

Herschel, C. 1888. *Apparatus for Measuring the Quantity of Water Flowing in a Pipe.* U.S. patent number US381,373. April 17.

Herschel, C. 1895. *The Venturi Meter.* Providence, RI: Builders Iron Foundry. Accessed August 7, 2017. http://www.ibiblio.org/kuphaldt/socratic/sinst/ebooks/The_Venturi_Meter.pdf.

McCabe, W.L., Smith, J.C., and Harriott, P. 2005. *Unit Operations of Chemical Engineering,* 7th ed. New York: McGraw-Hill.

Turnbull, H.W., (Ed.) 1959. *The Correspondence of Isaac Newton: 1661–1675*. 1st ed. London: Published for the Royal Society at the University Press. p. 416. Accessed August 7, 2017. https://en.wikipedia.org/wiki/Standing_on_the_shoulders_of_giants.

Wikipedia. 2016. *Clemens Herschel*. November 26. Accessed August 7, 2017. https://en.wikipedia.org/wiki/Clemens_Herschel.

Wikipedia. 2017. *Giovanni Battista Venturi*. April 28. Accessed August 7, 2017. https://en.wikipedia.org/wiki/Giovanni_Battista_Venturi.

5

A Simple Sharp-Edged Orifice Demonstration*

William Roy Penney, Shannon L. Servoss, Christa N. Hestekin, and Edgar C. Clausen

CONTENTS

5.1 Introduction

The most widely used flow measuring device in industry is the orifice meter. An orifice meter consists of an accurately machined and drilled plate, mounted between two flanges, with the hole most often set in the center of the pipe in which it is mounted. Pressure taps, located upstream and downstream of the plate, are used to measure the pressure differential, which is then used in calculating the fluid flow rate. The most common orifice plate is the sharp-edged orifice, which is usually beveled on the downstream side and has a sharp edge on the upstream side (Figure 5.1). The sharp-edged

---

* Reprinted from Penney, W.R. et al., *Proceedings of the ASEE Midwest Regional Conferences,* Copyright 2016, with permission from ASEE.

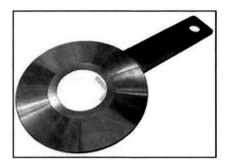

FIGURE 5.1
Sharp-edged orifice plates: upstream (left), downstream (right).

orifice is popular because of its low cost, simplicity, and small size, as well as the large amount of data available in describing its behavior and application (McCabe et al. 2005; Green and Perry 2008). Orifice meters containing sharp-edged orifice plates that are designed with standard dimensions yield discharge coefficients with errors of only 0.4%–0.8% (Green and Perry 2008; Reader-Harris 2015).

The objective of this paper is to document the construction and use of a simple experimental apparatus containing a sharp-edged orifice, followed by the presentation of results obtained when using this apparatus in a classroom demonstration to determine the coefficient of discharge (C_D) in a steady-state experiment. The experimentally determined C_D is then used to model the draining of the reservoir. This demonstration experiment is important educationally because:

- It requires students to correctly use a Bernoulli balance to model a steady-state system, and a Bernoulli balance and mass balance to model a transient system.
- It shows the students that this simple apparatus can yield a C_D that is very close to the accepted value of 0.61–0.65 (Alastal and Mousa 2015) for sharp-edged orifices.

5.2 Experimental

5.2.1 Apparatus

A photograph of the experimental apparatus (and students participating in a classroom demonstration) is shown in Figure 5.2. The apparatus consisted

FIGURE 5.2
Photograph of the experimental apparatus.

of a 10.2 cm (4 in) inside diameter, 61.6 cm (24.25 in)-long PVC pipe (0.63 cm or 0.25 in walls) containing the sharp-edged orifice at the bottom of the pipe, and attached upright to a metal support tripod. The PVC pipe had a sight glass tube (0.63 cm or 0.25 in clear PVC) attached to its side and connected to the inside of the reservoir at its bottom to observe liquid level in the pipe. A 64 L (17 gal) utility tub was used to collect water flowing from the pipe, and Erlenmeyer flasks and graduated cylinders were used to hold, feed, and collect water flowing in and out of the system. A stopwatch was used for timing the flow of water.

Figure 5.3 shows the PVC pipe from below, which shows the PVC plate containing the sharp-edged orifice (on the discharge side). The plate was fashioned from 1.3 cm (0.5 in) PVC, was 10.2 cm (4 in) in diameter and had a 6.35 mm (0.25 in) orifice at its center. As was noted earlier, the orifice must be properly designed and constructed with standard dimensions to minimize the error in C_D. In this case, the orifice plate was machined with a 30° downstream relief angle and a 0.5 mm (0.020 in) land, the minimum orifice wall thickness (Figure 5.4).

FIGURE 5.3
Photograph of the discharge of the orifice, as shown from the tube bottom.

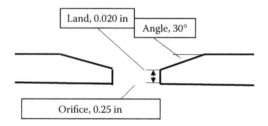

FIGURE 5.4
Dimensions of sharp-edged orifice.

5.2.2 Experimental Procedure

To begin an experiment, the entire apparatus (tripod, pipe, and tube with orifice) was placed in the utility tub. The orifice was plugged with a short length of Tygon® tubing, and the pipe was filled with water. Additional water was made available, as needed, for the experiment. Five students were recruited to perform the experiment.

To conduct a steady-state experiment, the Tygon plug was first removed from the orifice. Water was continuously poured into the reservoir to maintain the liquid level at the very top. Once steady state was reached, in a few seconds, the water flowing through the orifice was collected in a 2190 mL (volume completely filled to overflowing) Erlenmeyer flask. The time to fill the flask to overflowing was recorded. The plug was then replaced, and the experiment was repeated, as desired. To conduct a time-dependent experiment, the arrangement was essentially the same, except that the water was allowed to drain with time (as opposed to maintaining a constant level) after removing the plug. The water level indicated by the water level in the sight tube was recorded at periodic times from the start of the experiment when the plug was removed.

5.3 Experimental Data

Table 5.1 shows the raw experimental data for student-generated steady-state experiments. Five runs were made, collecting 2190 mL of water. Results from a time-dependent run are shown in Table 5.2, as the height of the water in the tank as a function of time.

TABLE 5.1

Experimental Data for the Steady-State Experiment[a]

Run	Collection Time, s
1	30.98
2	30.90
3	30.86
4	31.01
5	30.90

[a] The diameter of the pipe was 10.2 cm (4.0 in).

TABLE 5.2

Experimental Data for the Time-Dependent Runs

Time, s	Fluid Height, cm
0	60.96
7.24	55.88
13.59	50.80
20.28	45.72
27.53	40.64
35.17	35.56
42.75	30.48
52.15	25.40
61.82	20.32
73.11	15.24
86.36	10.16
103.1	5.08
133.6	0

5.4 Model Development

The basic Bernoulli Balance, with no work in the system and negligible friction losses, is described by Wilkes et al. (2006) and shown in Equation 5.1:

$$\frac{v_1^2}{2g} + z_1 + \frac{P_1}{\rho g} = \frac{v_2^2}{2g} + z_2 + \frac{P_2}{\rho g} \tag{5.1}$$

For application in this experiment, point 1 was selected as the fluid level in the pipe, and point 2 was selected as the location of the *vena contracta*, which is located about one-half of an orifice diameter from the orifice entrance (Calvert 2003). Since both ends of the tube were open to the atmosphere, $p_1 = p_2$. The velocity at the top of the liquid in the pipe, v_1, may be neglected, and the *vena contracta* is at zero height, so that $z_2 = 0$. With these simplifications, Equation 5.1 may be rearranged to solve for v_2, the velocity at the *vena contracta*, v_{vc}. This yields Equation 5.2:

$$v_2 = v_{vc} = \sqrt{2gz_1} \tag{5.2}$$

The C_D of the orifice may be described by Equation 5.3:

$$Q = C_D A_2 v_{vc} \tag{5.3}$$

where A_2 is the area of the orifice, equal to $\frac{\pi d_0^2}{4}$. Thus, C_D may be calculated by Equation 5.4:

$$C_D = \frac{Q}{A_2 \sqrt{2gz_1}} \tag{5.4}$$

for the steady-state system, where the volumetric flow rate is calculated as the volume of water collected, divided by the time of collection ($Q = v/t$).

In considering the time-dependent system, the simplified Bernoulli balance of Equation 5.2 must be combined with the mass balance, shown in Equation 5.5 as

$$\frac{dm}{dt} = m_1 - m_2 \tag{5.5}$$

For a draining tank, $m_1 = 0$, since there is no water flowing into the tank. Furthermore, $\frac{dm}{dt}$ may be written as $\rho A \frac{dh}{dt}$, and m may be written as $\rho v A$. Thus, Equation 5.5 becomes Equation 5.6, or

$$\rho A_1 \frac{dh}{dt} = -\rho v_2 A_2 \tag{5.6}$$

Combining Equations 5.2 and 5.6 yields Equation 5.7:

$$\frac{dh}{dt} = -\frac{A_2}{A_1}\sqrt{2gh} \tag{5.7}$$

Separating variables and integrating Equation 5.7 from $h = h_0$ at $t = 0$, and $h = h$ at $t = t$ yields, with rearrangement, Equation 5.8:

$$h = \left(\frac{C_D t A_2 \sqrt{2g}}{-2A_1} + \sqrt{h_0} \right)^2 \tag{5.8}$$

Finally, taking the square root of each side yields Equation 5.9:

$$\sqrt{h} = \frac{C_D t A_2 \sqrt{2g}}{-2A_1} + \sqrt{h_0} \tag{5.9}$$

Thus, a plot of versus t will yield a straight line, which is the usual method of presenting tank draining data.

5.5 Reduced Results and Discussion

5.5.1 Steady-State Results

The average volumetric flow rate was calculated from the experimental data in Table 5.1 ($Q = 7.085 \frac{m^3}{s}$) and combined with the geometrical variables ($A_2 = 3.17\text{E-}5 \text{ m}^2$, $z_1 = 24.375$ in [0.610 m]) in Equation 5.4 to yield a C_D of 0.641. Wilkes et al. (2006) note that the discharge coefficient should be about 0.63 for these operating conditions ($Re_o = 14{,}000$). This agrees well with the experimental value from the steady state runs of 0.641.

5.5.2 Time-Dependent Results

Figure 5.5 shows plots of versus t for the experimental data and the model prediction from Equation 5.9. The drain time predicted by the model agreed very well with the experimental data, within about 3%, except for the last data point.

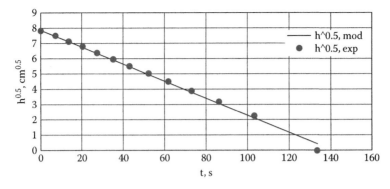

FIGURE 5.5
Plot of time-dependent experimental data with the model prediction of Equation 5.9.

5.6 Conclusions

1. The experiment is an excellent teaching tool because it shows students how a sharp-edged orifice must be designed and machined, and applies the Bernoulli and mass balances to reduce experimental data and develop a tank draining model.

2. Data reduction involves solution of a differential equation, which is excellent practice for the students.

3. The well-designed orifice yielded discharge coefficients that were almost identical to those described in the literature, with errors of about 3%.

4. This experiment meets all the requirements of a well-designed classroom experiment:

    • The apparatus is inexpensive

    • The experiments can be easily and quickly conducted in the classroom

    • Fundamental principles can be applied to model the experiment

    • Model development involves the solution of a differential equation and application of appropriate boundary conditions

    • The experimental results agree with literature data

    • The experimental data and model predictions are easily compared using linear plots; agreement is excellent

5.7 Nomenclature

Latin Symbols

A	Area, m^2
A_1	Area of the PVC pipe, 0.0081 m^2
A_2	Area of the orifice, 3.17×10^{-5} m^2
C_D	Orifice discharge coefficient, dimensionless
d_o	Diameter of orifice, 0.0064 m (0.25 in)
G	Gravitational constant, 9.8 $\frac{m}{s^2}$
H	Height of the liquid in the tank or pipe, m
h_o	Initial height of the liquid in the tank or pipe, m
M	Mass, kg
m_1	Mass of water entering the pipe, kg
$p_1 = p_2$	Pressures at top of liquid in the pipe and at exit of orifice, kPa
Q	Volumetric flow rate, $\frac{m^3}{s}$
T	Time, s
t_e	Time to empty the tank or pipe, s
V	Volume of water collected from apparatus, m^3
V	Velocity, $\frac{m}{s}$
v_1	Velocity at the top of the water in the pipe, $\frac{m}{s}$
$v_2 = v_{vc}$	Velocity leaving the orifice; velocity in the vena contracta, $\frac{m}{s}$
z_1	Height of water in the pipe, m
z_2	Height of water at the orifice, arbitrarily set at 0 m

Greek Symbols

ρ	Density of water, 1000 $\frac{kg}{m^3}$
μ	Viscosity of water, 0.1 $\frac{kg}{ms}$

Dimensionless Groups

Re_o	Reynold's number through the orifice, $\frac{d_o v_2 \rho}{\mu}$, dimensionless

Acknowledgment

The authors acknowledge the skills of Mr. George Fordyce of the University of Arkansas, Ralph E. Martin Department of Chemical Engineering, in machining the orifice and fabricating the apparatus to required specifications.

References

Alastal, K.M. and Mousa, E.M.Y. 2015. *Fluid Mechanics and Hydraulics Lab Manual. Experiment 5: Flow through Small Orifices.* Accessed August 7, 2017. http://site.iugaza.edu.ps/mymousa/files/Experiment-5.pdf.

Calvert, J.B. 2003. *Coefficient of Discharge.* June 15. Accessed August 7, 2017. http://mysite.du.edu/~jcalvert/tech/fluids/orifice.htm.

Green, D.W. and Perry, R.H. 2008. *Perry's Chemical Engineers' Handbook,* 8th ed. New York: McGraw-Hill.

McCabe, W.L., Smith, J.C., and Harriott, P. 2005. *Unit Operations of Chemical Engineering,* 7th ed. New York: McGraw-Hill.

Reader-Harris, M. 2015. *Orifice Plates and Venturi Tubes.* Switzerland: Springer International.

Wilkes, J.O., Birmingham, S.G., Kirby, B.J., and Cheng, C.Y. 2006. *Fluid Mechanics for Chemical Engineers,* 2nd ed. Boston, MA: Pearson Education.

6

Depressurization of an Air Tank[*]

Martin A. Christie, John A. Dominick III,
William Roy Penney, and Edgar C. Clausen

CONTENTS

6.1 Introduction

On October 13, 1998, a plug of sludge-like material caused an explosion and fire inside a 11.4 m^3 (3000 gal) Hastelloy reactor, as part of the linear alkyl benzene process at the Condea Vista plant in Baltimore, Maryland (Reza et al. 2002). The explosion fueled a fire that took about two hours to extinguish. Chemical processing equipment must be designed so that such a catastrophe never happens. Leung (1992) explains how to implement pressure relief by using *capability for latent heat of cooling via boiling* to generate vapor within the reactor, which is vented in a controlled manner from the vessel to maintain the vessel pressure within safe levels. In order to accomplish

---

[*] Adapted from Christie, M.A. et al., *Proceedings of the ASEE Midwest Regional Conference*, Copyright 2017, with permission from ASEE.

the mathematical modeling required to design such a system, the following must be accomplished:

1. Perform a mass and heat balance on the reactor contents, which results in differential equations.
2. Determine the vapor venting rate from the vessel through pipe, pipe fittings, and orifices.
3. Solve the differential equations required to predict pressure versus time within the reactor.

In this depressurization experiment, the students practiced the following to model the depressurization:

1. Perform a mass balance on the vessel contents.
2. Calculate the vapor venting rate under non-critical and critical flow conditions.
3. Develop a differential equation that must be solved to determine pressure versus time in the vessel.
4. Write a computer program to solve the differential equation.

The computer program must include the logic to handle both sub-critical and critical flow through the venting orifice. The objective of this paper is to describe a simple experiment for the depressurization of a 0.042 m³ (11 gal) air tank through a 1.32 mm (0.052 in) sharp-edged orifice. Experimental data of pressure versus time were compared to computer model predictions. The computer model used an Euler integration to solve the governing differential equation.

6.2 Experimental

6.2.1 Apparatus

A photograph of the experimental apparatus is shown in Figure 6.1. The apparatus consisted of a 0.042 m³ (11 gal) Campbell–Hausfeld carbon steel air tank with a maximum pressure rating of 125 psig (8.5 atm gauge) and a calculated actual volume of 0.044 m³ (11.6 gal). The tank was equipped with a 6.4 mm ($\frac{1}{4}$ in) valve to initiate depressurization, a 0–160 psig (0–10.9 atm gauge) pressure gauge, and a 6.4 mm ($\frac{1}{4}$ in) brass pipe plug. To form the sharp-edged orifice, a 6.4 mm ($\frac{1}{4}$ in) brass pipe plug was drilled from both sides, a 7.94 mm $\left(\frac{5}{16}\text{ in}\right)$ square-bottom drill was used to form the inlet cavity, and a

FIGURE 6.1
Photograph of the air tank and respective attachments for venting data collection.

FIGURE 6.2
Photograph of the brass plug and orifice, shown from the outlet side.

6.4 mm ($\frac{1}{4}$ in) partially square-tip drill was used to form the outlet cavity. The drilled holes formed a 0.25 mm (0.01 in) plate about midway through the pipe plug. This plate was drilled in its center with a 1.32 mm (0.052 in) diameter drill to complete the orifice. Figure 6.2 shows a photograph of the brass fitting.

6.2.2 Experimental Procedure

Prior to experimentation, the tank was checked for any defects. The tank was then pressurized to 5.4 atm absolute (65 psig) with shop air before moving the apparatus to the classroom. To begin the experiment, the ball valve was opened fully, and pressure was then measured and recorded as a function of time. The time was initially recorded for each 0.34 atm (5 psi) reduction in pressure and later for each 0.17 atm (2.5 psi) reduction in pressure. The experiment ended when the tank reached a pressure of 1.5 atm absolute (7.5 psig).

6.2.3 Safety Concerns

Proper safety equipment for this experiment includes the wearing of safety goggles, long pants, and protective gloves. Prior to tank venting, the orifice must be free of obstruction, and the path of the pressurized air must be clear to avoid damage to students and the surroundings.

6.3 Experimental Data

Table 6.1 presents the experimental data of tank pressure readings and the time required to reach the pre-selected tank pressure.

TABLE 6.1

Experimental Data—Tank Pressure
versus Time

Time (s)	Tank Pressure (psig)
0	65.0
10.32	60.0
18.94	55.0
30.45	50.0
48.39	45.0
66.29	40.0
93.96	35.0
114.44	30.0
130.65	27.5
144.75	25.0
158.52	22.5
170.39	20.0
181.26	17.5
197.84	15.0
223.46	12.5
245.45	10.0
263.47	7.5

6.4 Model Development

Equation 6.1 is obtained in performing a mass balance on the tank. Since the tank has no inlet streams or generation of air, Equation 6.1 reduces to Equation 6.2:

$$m_i - m_o + m_g = m_a \qquad (6.1)$$

$$-m_o = m_a \qquad (6.2)$$

The mass accumulated within the tank, m_a, may be expressed as a differential change in the mass of air in the tank using the continuity equation, shown in Equation 6.3

$$m_a = \frac{dM}{dt} = \frac{d(\rho_t V)}{dt} \qquad (6.3)$$

The density of the gas can be expressed in Equation 6.4 using the ideal gas equation

$$\rho_t = \frac{P_t(MW)}{RT} \qquad (6.4)$$

Expanding Equation 6.3 to also include the ideal gas equation yields Equation 6.5 for a constant T:

Finally, the differential pressure change as a function of changing mass flow rate is found by a rearrangement of Equation 6.5, resulting in Equation 6.6:

$$m_a = \frac{d\left(\dfrac{P_t(MW)V}{RT}\right)}{dt} = -m_o \qquad (6.5)$$

$$\frac{dP_t}{dt} = -\frac{m_o RT}{V(MW)} \qquad (6.6)$$

The mass velocity of gas leaving a reservoir, through a frictionless ideal nozzle, may be calculated from Equation 6.7 (McCabe et al. 2005, p. 143, Equation 6.30):

$$G = \sqrt{\frac{2\gamma \rho_t P_t}{\gamma - 1}} \left(\frac{P_{vc}}{P_t}\right)^{\frac{1}{\gamma}} \sqrt{1 - \left(\frac{P_{vc}}{P_t}\right)^{\left[1 - \left(\frac{1}{\gamma}\right)\right]}} \qquad (6.7)$$

The constant γ is the specific heat ratio, which is 1.4 for air.

In Equation 6.7, the gas density within the tank, ρ_t, and pressure at the vena contracta, P_{vc}, were found using Equations 6.4 and 6.8, respectively:

$$P_{vc} = P_t(r_c) \tag{6.8}$$

The parameter, r_c, the critical pressure ratio, is calculated by Equation 6.9 (McCabe et al. 2005, p. 142, Equation 6.29):

$$r_c = \left(\frac{2}{\gamma+1} \right)^{\frac{1}{1-\frac{1}{\gamma}}} \tag{6.9}$$

For air, $r_c = 0.53$. For $\frac{P_a}{P_t} < 0.53$, where sonic velocity occurs, P_{vc} in Equation 6.7 was set equal to the vena contracta pressure $= 0.53 \, P_t$; for $\frac{P_a}{P_t} > 0.53$, P_{vc} was set equal to the downstream pressure, that is, atmospheric pressure.

The mass flow from the tank was calculated in Equation 6.10 using the ideal mass velocity (G), the cross-sectional area of the orifice ($A_o = \frac{\pi d^2}{4}$), a dimensionless expansion factor (Y), and the orifice coefficient, C_d

$$m_o = C_d Y G A o \tag{6.10}$$

The orifice coefficient was found using a linear regression of C_d versus $1 - P_r$ data from Linfield (2015, pp. 69–70, Figures 3.12 and 3.13). P_r is the pressure ratio of atmospheric pressure to absolute tank pressure, $\frac{P_{atm}}{P_t}$. The regression results in a fourth-order equation, shown in Equation 6.11:

$$C_d = 0.6219 + 0.0686\left(1 - P_r\right) + 0.7955\left(1 - P_r\right)^2 - 0.9285\left(1 - P_r\right)^3$$
$$+ 0.2914\left(1 - P_r\right)^4 \tag{6.11}$$

According to McCabe et al. (2005, p. 232, Equation 8.33), if the critical pressure ratio, r_c, is less than 0.53 for air, the gas flow is sonic and $Y = 1$. For other r_c values, the expansion factor was computed as noted in Equation 6.12:

$$Y = 1 - \frac{0.41 + 0.35\beta^4}{\gamma}\left(1 - \frac{P_{atm}}{P_t}\right) \tag{6.12}$$

To determine P as a function of time, the mass flow from the tank is first found using Equations 6.7 through 6.12. An Euler integration was then used to solve the differential equation in Equation 6.5.

6.5 Results and Discussion

Figure 6.3 presents a plot of the experimentally measured tank pressures and the model-predicted pressures versus time. There is minimal deviation in the mathematical model from the experimental data, indicating that the model is adequate. The mathematical model predicted slightly higher pressures (by an average of 14%) than the experimental data from 0 to 75 s, while the predicted values were only slightly lower than the experimental values for the time range of 100–175 s. The model matched the experimental data very well for times greater than 175 s. Due to the simplicity of the experiment, the system required minimal user interaction. Thus, errors were minimal.

6.6 Conclusions

1. The mathematical model predictions and results fit the experimental data well throughout the depressurization time. The maximum variance of the model from the data was 14% for the initial time step of 0–75 s.

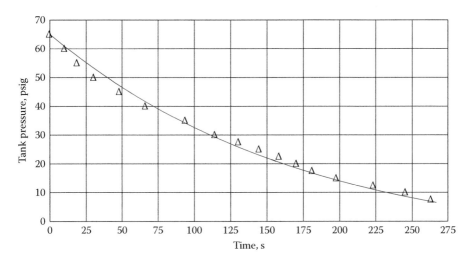

FIGURE 6.3
Experimental and model results from the tank depressurization experiment.

2. Errors in the experiment were most likely due to human error as a result of the lag between reading the pressure and corresponding time. This error has been minimized in an upgrade of the experimental system by the addition of a Measurement Computing data acquisition device, USB-TC-AI, driven by a 2 amp/12 volt source and connected to the USB port of a Dell Latitude E 5510 laptop computer. Omega TracerDAQPro software can be used to display and analyze the data.

6.7 Nomenclature

Latin Symbols

A_o Cross-sectional area of the orifice, m^2

C_d Orifice discharge coefficient, dimensionless

d Orifice diameter, m

$\dfrac{dM}{dt}$ Change of mass in the tank with respect to time, $\frac{kg}{s}$

$\dfrac{dP_t}{dt}$ Change in tank pressure with respect to time, $\frac{Pa}{s}$

G Mass velocity, $\frac{kg}{m^2 s}$

m_a Mass accumulation within the tank, $\frac{kg}{s}$

m_g Mass generation within the tank, $\frac{kg}{s}$

m_i Mass entering the tank, $\frac{kg}{s}$

m_o Mass exiting the tank, $\frac{kg}{s}$

MW Molecular weight of air, $\frac{kg}{kgmol}$

P_{atm} Atmospheric pressure, P_a

P_r Atmospheric-to-tank pressure ratio, $\frac{P_a}{P_t}$

P_t Tank pressure, Pa

P_{vc} Pressure at the vena contracta, P_a

R Gas constant, $\frac{cm^3 kPa}{gmol\ K}$

r_C Critical pressure ratio

T Tank temperature, K

V Tank volume, m^3

Y Gas expansion factor

Greek Symbols

β   Diameter ratio, orifice diameter/cavity diameter

γ   Specific heat ratio, $\frac{C_p}{C_v}$

ρ_t Air density, $\frac{kg}{m^3}$

Acknowledgment

The authors acknowledge the efforts of Mr. George Fordyce of the University of Arkansas, Ralph E. Martin Department of Chemical Engineering, in machining the orifice and constructing the experimental apparatus.

References

Leung, J. C. 1992. Venting of runaway Reactions with gas generation, *AIChE Journal* 38(5) 723–732.

Linfield, K. W. 2000. A study of the discharge coefficient of jets from angled slots and conical orifices, PhD Dissertation, University of Toronto, Toronto, ON, p. 70. Accessed August 8, 2017. http://www.collectionscanada.gc.ca/obj/s4/f2/dsk1/tape3/PQDD_0028/NQ49816.pdf.

McCabe, W. L., Smith, J.C., and Harriott, P. 2005. *Unit Operations of Chemical Engineering*, 6th ed., p. 143. New York: McGraw-Hill.

Reza, A., Kemal, A., and P. E. Markey. 2002. Runaway reactions in aluminum, aluminum chloride, HCl and steam: An investigation of the 1998 CONDA Vista explosion in Maryland. *Process Safety Progress* 21(3) 261–267.

7

Draining an Upper Tank into a Lower Tank: Experimental and Modeling Studies

William Roy Penney and Edgar C. Clausen

CONTENTS

7.1 Introduction

Tank draining is a very important topic in engineering practice, as evidenced by a Google Search for *calculating tank drainage* that gives about 2,940,000 results and a search on YouTube of *tank drainage* that gives about 256,000 results. Problems involving the draining of liquid from tanks, and from tanks into other tanks, can be solved by using a combination of mass balances and Bernoulli balances. This experiment is an excellent learning tool for applying mass and Bernoulli balances and for solving the differential equations that are included in the modeling of tank draining problems.

The purpose of this experiment was to

- Conduct a tank draining experiment where liquid levels were measured as one tank drained into another.

- Model the experiment to predict the liquid levels in both tanks as a function of drain time.
- Compare model predictions with the experimental data.

7.2 Experimental

7.2.1 Equipment List

The following equipment was required to carry out the experiment:

- Transparent acrylic reservoir (5.1 cm [2 in] diameter × 40 cm [15 $\frac{3}{4}$ in] high):
  - A 9.5 mm ($\frac{3}{8}$ in) hole was drilled at the bottom to enable a 6.4 mm ($\frac{1}{4}$ in) inside diameter silicone tube to be inserted as a re-entrant tube.
  - A measuring tape (graduated in inches) was taped to the back of the tube so that the liquid level in the tube could be observed and recorded.
- Opaque Schedule 40 PVC tube (3.5 cm [1 $\frac{3}{8}$ in] inside diameter, 4.21 cm [1 $\frac{3}{8}$ in] outside diameter, 86.4 cm [34 in] high)
- 2.5 cm (1 in) Schedule 40 PVC pipe, 194 cm (76.5 in) long, used as a standby insertion into the base tube
- 3.2 cm (1 $\frac{1}{4}$ in) Schedule 40 PVC pipe, 30.4 cm (12 in) long
- Silicone drain tube (6.4 mm [$\frac{1}{4}$ in] inside diameter, 9.5 mm [$\frac{3}{8}$ in] outside diameter, 1.7 m high), manufactured by Generic
- Wood base, 3.8 × 28.6 × 53.3 cm (1 $\frac{1}{2}$ in × 11 $\frac{1}{4}$ in × 21 in)
- 2 L Erlenmeyer flask, polypropylene
- Angle bracket, 5.1 × 5.1 cm × 3.1 mm thick (2 × 2 × $\frac{1}{8}$ in thick)
- Rubber stopper, 3.5 cm (1 $\frac{3}{8}$ in), at mid-length, used as a stopper for the 86.3 cm long, 3.2 cm (34 in long, 1 $\frac{1}{4}$ in) PVC bottom reservoir
- Meter stick
- Clear, transparent tape

7.2.2 Experimental Apparatus

Figure 7.1 is a photograph of the entire apparatus. It was constructed as follows:

1. A stand was made by attaching the 3.2 cm (1 $\frac{1}{4}$ in), 30.4 cm (12 in) long Schedule 40 PVC pipe to a 3.8 × 28.6 × 53.3 cm (1 $\frac{1}{2}$ × 11 $\frac{1}{4}$ × 21 in) wood base by using an angle bracket and a hose clamp.

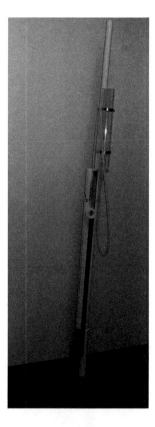

FIGURE 7.1
The experimental apparatus.

2. One end of the opaque Schedule 40 PVC tube was closed with a rubber stopper.

3. The silicone tube was inserted into the clearance hole in the bottom of the transparent acrylic reservoir.

4. The 3.2 cm ($1\frac{1}{4}$ in) × 30.4 cm (12 in) long Schedule 40 PVC pipe was taped (using clear transparent tape) to the 2.5 cm (1 in) × 194 cm (76.5 in) long Schedule 40 PVC pipe, with the stoppered end 30.5 cm (12 in) from the end of the stand pole.

5. The acrylic reservoir was taped to the stand pole with its closed end positioned at the top of the 3.2 cm ($1\frac{1}{4}$ in) PVC pipe reservoir.

6. The stand pole was inserted into the 3.2 cm ($1\frac{1}{4}$ in) × 30.4 cm (12 in) long pipe, which was secured to the base.

7.2.3 Experimental Procedure

1. Assemble the apparatus as shown in Figures 7.1 and 7.2.

FIGURE 7.2
The attached clear acrylic reservoir, the PVC reservoir, the silicone drain tube, the top reservoir scale, and the meter stick.

2. Pinch the silicone tubing just below the outlet of the top reservoir so the tube does not leak water.

3. Hold the apparatus upright and pour water in to the upper (clear) acrylic reservoir to the very top.

4. Place the apparatus support pole into the 3.2 cm ($1\frac{1}{4}$ in) × 30.4 cm (12 in) long PVC pipe attached to the wood base.

5. Feed the tube into the lower reservoir until it hits the bottom of the capped PVC pipe.

6. Release the pinch on the silicone tubing. This starts the experiment.

7. Start the stopwatch when the pinch is released.

8. Use the lap feature on the stopwatch to time the level decrease in the top reservoir as the water level in this reservoir passes predetermined levels.

7.2.4 Safety

Wear safety goggles, closed-toe shoes, and pants throughout the experiment.

7.3 Experimental Data

Table 7.1 displays the experimental data, taken as two runs during the experiment.

7.4 Reduction of Experimental Data

The overall mass balance for any system is given in Equation 7.1 as:

$$m_i + m_o + m_g = m_a \tag{7.1}$$

TABLE 7.1

Experimental Data

Water Height, in Clear PVC (Tank 1)	Time, s Experiment 1	Change in Time, s Experiment 2	Time, s Experiment 2	Change in Time, s Experiment 2
14.5	0	0	0	0
13.5	0.76	0.76	1.0	1.0
12.5	1.99	1.23	2.04	1.04
11.5	3.25	1.26	3.52	1.48
10.5	4.51	1.26	4.83	1.31
9.5	5.92	1.41	6.26	1.43
8.5	7.45	1.53	7.67	1.41
7.5	8.99	1.54	9.21	1.54
6.5	10.45	1.46	10.82	1.61
5.5	12.58	2.13	12.87	2.05
4.5	14.49	1.91	14.86	1.99
3.5	16.34	1.85	16.71	1.85
2.5	18.97	2.63	19.36	2.65
1.5	22.08	3.11	22.33	2.97
0.5	26.04	3.96	26.3	3.97
0	29.4	3.36	29.79	3.49

The mass in the upper and lower reservoirs can be expressed as in Equations 7.2 and 7.3:

$$M_u = \rho\left(\frac{\pi}{4}\right)D_u^2 h_u \tag{7.2}$$

$$M_l = \rho\left(\frac{\pi}{4}\right)D_l^2 h_l \tag{7.3}$$

The accumulation terms in both tanks are shown in Equations 7.4 and 7.5:

$$\frac{dM_u}{dt} = \rho\left(\frac{\pi}{4}\right)D_u^2\left(-\left[\frac{dh_u}{dt}\right]\right) \tag{7.4}$$

$$\frac{dM_l}{dt} = \rho\left(\frac{\pi}{4}\right)D_l^2\left(\frac{dh_l}{dt}\right) \tag{7.5}$$

All the water leaving the upper tank flows into the lower tank (i.e., $\frac{dM_u}{dt} = \frac{dM_l}{dt}$), and those mass flow rates are equal to the mass flow rate through the drain tube (i.e., $m_t = v_t\frac{\pi}{4D_t^2}$), thus, as is noted in Equation 7.6:

$$-\left[\frac{dh_u}{dt}\right] = \left(\frac{D_l^2}{D_u^2}\right)\left(\frac{dh_l}{dt}\right) = v_t\left(\frac{D_t^2}{D_u^2}\right) \tag{7.6}$$

A Bernoulli balance must be applied to determine the velocity in the drain tube. The following premises are applicable to perform a proper Bernoulli balance:

1. The system is defined as the water in both reservoirs.
2. The inlet to the system is the water-free surface in the upper tank, defined as point 1.
3. The exit of the system is the water-free surface in the lower tank, defined as point 2.
4. The pressure at the inlet and outlet are equal (i.e., $P_1 = P_2$) because they are atmospheric.
5. The free surface velocities are negligible relative to the velocity in the tube; consequently, $v_1 \approx 0$ and $v_2 \approx 0$; for purposes of including the velocity terms in the Bernoulli balance.

With these premises, the Bernoulli balance (McCabe et al. 2005) for the defined system is shown in Equation 7.7:

$$z_1 = z_2 + F_t = z_2 + \frac{v_t^2}{2g}\left(f_D \frac{L_t}{D_t} + \sum k \right) \tag{7.7}$$

If we define $\Delta z = z_1 - z_2$, then the velocity in the drain tube can be obtained from Equation 7.8:

$$v_t = \sqrt{\frac{2g\,\Delta z}{\left[f\dfrac{L_t}{D_t} + \sum k \right]}} \tag{7.8}$$

The friction factor can be determined by use of the Wood Equation (Eng-Tips 2017), for Re > 4000 and any $\frac{e}{D}$, shown in Equation 7.9:

$$f = 0.94\left(\frac{e}{D}\right)^{0.225} + 0.53\left(\frac{e}{D}\right) + 88\left(\frac{e}{D}\right)^{0.44} \times \mathrm{Re}^a \tag{7.9}$$

$$a = -1.62\left(\frac{e}{D}\right)^{0.134}$$

The procedure utilized to solve the differential equations of the model is as follows:

1. Define a reasonable time increment to perform an Euler integration, perhaps 1 s for the 30 s drain time.
2. Determine v_t from Equation 7.8—$v_t = \sqrt{2g\Delta z / \left[f(L_t/D_t) + \Sigma k \right]}$ using the friction factor from the previous iteration.
3. Solve Equation 7.6 for $H_{u,i} = H_{u,i-1} + \left(\Delta H_u / \Delta t \right)$ and $dh_l/dt \approx \Delta H_{l,j}/\Delta t$ and then determine $H_{u,i}$ and $H_{l,i}$ as follows in Equation 7.10:

$$H_{u,i} = H_{u,i-1} + \left(\tfrac{\Delta H_u}{\Delta t}\right)\Delta t \text{ and } H_{l,i} = H_{l,i-1} + \left(\tfrac{\Delta H_l}{\Delta t}\right)\Delta t \tag{7.10}$$

7.5 Reduced Results

Figure 7.3 shows a plot of the model prediction and the experimental data.

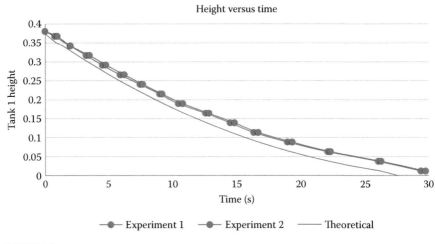

FIGURE 7.3
Plots of the model prediction and the experimental data.

7.6 Discussion of Results, Conclusions, and Recommendations

The model-predicted drain time is about 10% less than the experimental drain time. Deviations of model predictions from experimental data can result from:

1. Errors in the tube diameter measurement. In Equations 7.4 and 7.5, the tube diameter appears to the second power; thus, a 1% error in the tube diameter will result in a 2% error in the calculated discharge tube velocity. The tube inside diameter was measured by inserting a twist drill shank into the tube; a more accurate measurement could result in a more accurate model.

2. There are errors in the measurement of the time for the level in the upper tank to reach the next tick mark. A camera recording the reading on a stopwatch and the tick marks would give greater accuracy.

3. The Wood equation is accurate to at most 5%; there are at least six other equations shown by Eng-Tips (2017) that might be more accurate and give a better fit between model predictions and experimental data.

4. The minor loss coefficients were: re-entrant opening to the tube = 0.8 and tube exit = 1. These were taken from the literature; their appropriateness should be checked for future experiments.

5. Only moderate efforts were made to keep the experimental apparatus exactly vertical. Future experiments should be designed to keep the apparatus vertical.

6. The experiment was only conducted once, with two students doing timing. Duplicate runs could be used to reduce the errors arising from experimental errors.

7.7 Nomenclature

Latin Letters

$dH_l \approx \Delta H_l$	Change in liquid level in the lower tank during a time increment Δt, m
$dH_u \approx \Delta H_u$	Change in liquid level in the upper tank during a time increment Δt, m
D_l	Inside diameter of the lower tank, m
D_t	Inside diameter of the upper tank drain, m
$dt \approx \Delta H_l$	Time increment for performing the Euler integration, s
D_u	Inside diameter of the upper tank, m
e	Roughness height for the drain tube, m
f_D	Darcy friction factor for pipe flow, unitless
g	Earth gravitational constant, $\frac{m}{s^2}$
F_t	Head loss arising from frictional flow in the drain tube, m
h_u	Height of water in the upper tank—free surface to tank bottom, m
h_l	Height of water in the lower tank—free surface to tank bottom, m
H_u	Height water-free surface in the upper tank above the datum at the laboratory floor, m
H_l	Height water-free surface in the lower tank above the datum at the laboratory floor, m
k	Minor loss coefficient for head loss in pipe fittings, including inlet and outlet, unitless
L_t	Length of drain tube, m
m_a	Mass accumulated in the system, kg
m_g	Mass rate generated, kg
m_i	Mass into the system, kg
m_o	Mass out of the system, kg
M_u	Mass in the upper tank, kg
M_l	Mass in the lower tank, kg

P_1	Pressure at the top of the upper tank, Pa
P_2	Pressure at the top of the water in the lower tank, Pa
v_l	Velocity of the free surface in the lower tank, $\frac{m}{s}$
v_t	Velocity of the water in the drain tube, $\frac{m}{s}$
v_u	Velocity of the free surface in the upper tank, $\frac{m}{s}$
z_1	Height of the free surface in the upper tank above the laboratory floor, m
z_2	Height of the free surface in the lower tank above the laboratory floor, m
$\Delta z = z_1 - z_2$	*Difference*: Height of the free surface in the upper tank minus height of the surface in the lower tank, m

Greek Symbols

ρ	Density of water, $\frac{kg}{m^3}$
μ	Viscosity of water, $\frac{kg}{m\,s}$

Acknowledgment

This chapter contains results from laboratory reports prepared by former University of Arkansas, Ralph E. Martin Department of Chemical Engineering, students including Alexander L. Branstetter. The authors are extremely grateful for the hard work and dedication of these University of Arkansas graduates.

References

McCabe, W.L., Smith, J.C., and Harriott, P. 2005. *Unit Operations of Chemical Engineering,* 7th ed., p. 125. New York: McGraw-Hill.

Eng-Tips. 2017. *Pipelines, Piping and Fluid Mechanics Engineering FAQ.* Accessed August 20, 2017. http://www.eng-tips.com/faqs.cfm?fid=1236.

8

Determining the Pump Curve and the Efficiency for a Bomba BP-50 Regenerative Centrifugal Pump

William Roy Penney and Edgar C. Clausen

CONTENTS

8.1 Introduction

Pump curves are the basic tools needed to select and design pumps for use in pumping systems. The typical centrifugal pump curve presents the following information (Schroeder 2014):

1. Developed head versus flow for a selected pump speed for several impeller sizes
2. Net Positive Suction head (NPSH) required
3. Horsepower for water as the pumped fluid
4. Pump efficiency curves for the various selected impeller sizes

A typical pump curve is shown in Figure 8.1.

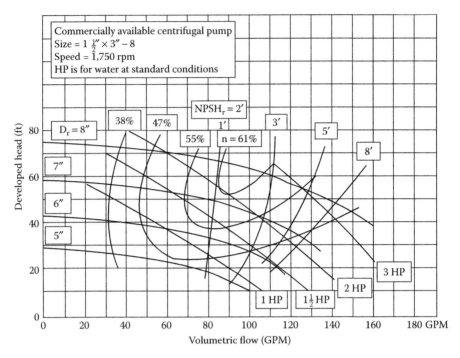

FIGURE 8.1
Pump curves for a typical commercially available centrifugal pump.

For the pump selected for this experiment, a flow curve and an overall efficiency curve were developed. The manufacturer does not give the NPSH requirements for this pump and they were not determined during this investigation.

8.2 Experimental

8.2.1 Experimental Equipment

A listing of the experimental equipment is shown in Figure 8.2, including the pump, fittings, pressure gauge, wattmeter, and auxiliary equipment.

8.2.2 Assembly of the Experimental Apparatus

To begin assembly, a 2.5 cm (1 in) short PVC nipple was inserted and tightened into the discharge port of the pump. The nipple was then inserted into the hole in the feed reservoir. Silicone caulking was applied inside the pail around the junction of the pail wall and the nipple; then, a 2.5 cm (1 in) PVC

1. Bomb DP-50 centrifugal pump manufactured
   by DOMOSA.

The pump is listed on eBay under
"$\frac{1}{2}$ HP electric peripheral clear water pump."
As of 8-23-17 the list price was $30.95
It is a new pump.
The shipping is free.

2. Three 1" PVC pipe nipples.
3. One 1" pipe tee with a $\frac{1}{2}$" branch.
4. A 0–100 psi Ashcroft test guage.
5. Two 5 gal polyethylene pails
   - A 1 $\frac{1}{4}$" hole was drilled into the side of the feed pail
     to accommodate the 1" nipple, which was installed
     into the discharge port of the pump.
6. One 1" PVC ball valve.
7. A 4' long section of 1" Tygon© tubing.
8. A 0–360 lb electronimc platform scale.
9. A 1' × 2' piece of $\frac{3}{4}$" plywood.
10. A 2' × 3' piece of $\frac{1}{2}$" plywood.
11. Wattmeter – KILL A WATT – Model P4460.01

FIGURE 8.2
A composite covering the equipment required to construct the experimental apparatus.

pipe coupling was screwed tightly over the nipple, inside the tank, so that
there was a tight seal around the suction line where it penetrated the pail
wall. The discharge line was assembled as follows:

1. A 2.5 cm (1 in) PVC nipple was screwed into the pump discharge.
2. A 2.5 cm (1 in) galvanized tee with a 1.3 cm ($\frac{1}{2}$ in) branch was screwed
   to the nipple.
3. A 1.3–0.6 cm ($\frac{1}{2}$ to $\frac{1}{4}$ in) stainless steel bushing was inserted into the
   tee branch.
4. The pressure gauge was installed into the bushing.
5. A 2.5 cm (1 in) PVC nipple was screwed into the tee.
6. A 2.5 cm (1 in) PVC ball valve was attached to the tee.

FIGURE 8.3
Photograph of the experimental apparatus.

7. A 2.5 cm (1 in) PVC hose barb was screwed into the outlet of the ball valve.

8. A 1.2 m (4 ft) length of 2.5 cm (1 in) Tygon® tubing was attached to the hose barb with a hose clamp.

9. The pump and feed reservoir were attached to a wood base with screws as shown in Figure 8.3.

10. The pump electrical cord was connected to an electrical outlet through a power strip and the wattmeter.

8.2.3 Experimental Procedure

1. The feed reservoir was filled with water to within about 5.1 cm (2 in) of its top.

2. The ball valve was placed in the fully open position.

3. The pump was turned on by the switch on the power strip.

4. The pressure reading on the pressure gauge was recorded.

5. The water weighing pail was tared on the scale and placed as close to the feed reservoir as possible.

6. The Tygon® discharge tube was lifted from the water in the feed reservoir and placed quickly into the weighing pail; simultaneously, the stopwatch was started.

7. After about 11.4 L (3 gal) of water were collected in the weighing pail, the discharge tube was quickly moved back into the feed pail; simultaneously, the stopwatch was stopped.

8. The weighing pail and water were placed on the scale.

9. The weight of the water and the collection time were recorded.

10. The ball valve was then closed to increase the pump discharge pressure to the next pressure increment, and steps 4 through 9 were repeated until the last pressure increment occurred when the discharge valve was fully closed.

8.3 Data Reduction

The use of the Bernoulli balance (McCabe et al. 2005), shown in Equation 8.1, was essential to reduce the experimental data to produce curves of head versus flow and overall efficiency versus flow.

$$\frac{p_a}{\rho} + gZ_a + \frac{V_a^2}{2} = \frac{p_b}{\rho} + gZ_b + \frac{V_b^2}{2} + h_f \tag{8.1}$$

The relationship between volume flow rate and mass flow rate and head developed to determine pumping power (Milnes 2017) must also be used. This is shown in Equation 8.2:

$$P_p = \frac{g\rho QH}{\eta} = \frac{gMH}{\eta} \tag{8.2}$$

The mass flow rate is determined from the measured weight of water for the time period that the water discharge is diverted into the weighing pail, shown in Equation 8.3:

$$M = \frac{W}{t} \tag{8.3}$$

The head developed by the pump is determined by applying the Bernoulli balance twice.

First the pressure, relative to atmospheric, in the pump suction line must be determined. An application of the Bernoulli balance from the free water surface in the feed pail to the pump suction inlet line gives a relationship pressure at the pump suction, shown in Equation 8.4:

$$\frac{p_a}{\rho} + gZ_a + \frac{V_a^{\,2}}{2} = \frac{p_b}{\rho} + gZ_b + \frac{V_b^{\,2}}{2} \tag{8.4}$$

In Equation 8.4, p_a is atmospheric; thus, it is zero relative to atmospheric; V_a is the feed pail free surface velocity, which is very close to 0; V_b is the velocity in the 2.5 cm (1 in) inlet line, which is a maximum of 1.3 $\frac{m}{s}$, which gives a velocity head of about 0.1 m; thus, the inlet line velocity head can be neglected; the elevation datum is taken as the elevation of the pump suction line, so $Z_b = 0$; consequently, Equation 8.4 becomes Equation 8.5:

$$\frac{p_b}{\rho} = gZ_a = \frac{p_{il}}{\rho} \tag{8.5}$$

The head produced by the pump can be determined by performing a Bernoulli balance between the pump suction line and the pressure gauge just before the throttling valve. The equation for the head produced by the pump is shown in Equation 8.6:

$$\frac{p_{il}}{\rho} + gZ_{il} + \frac{V_{il}^{\,2}}{2} = \frac{p_b}{\rho} + gZ_b + \frac{V_{il}^{\,2}}{2} = \frac{p_{pg}}{\rho} + gZ_{pg} + \frac{V_{pf}^{\,2}}{2} + H_p \tag{8.6}$$

The velocity head terms are deemed negligible and $Z_{il} = Z_b = 0$; thus, Equation 8.6 becomes Equation 8.7:

$$H_p = \frac{(p_{pg} - p_{il})}{g\rho} + g(Z_{il} - Z_{pg}) = \frac{p_{pg}}{g\rho} - gZ_a + gZ_{pg} = \frac{p_{pg}}{g\rho} - g(Z_a + Z_{pg}) \tag{8.7}$$

For this apparatus, $Z_a = 30.5$ cm (12 in) and $Z_{pg} = 8.9$ cm (3.5 in).

The minimum power required to move the fluid with a 100% efficient prime mover is equal to gMH_p. The electrical power input to the electric motor was measured with a wattmeter. The overall pumping efficiency is given by Equation 8.8:

$$\eta = 100 \left(\frac{P_t}{P_m} \right) = 100 \left(\frac{gMH_p}{P_m} \right) \tag{8.8}$$

8.4 More about the Regenerative Pump

Regenerative centrifugal pumps will produce high heads at reasonable flows. The high heads produced by the pump result from the vanes machined on each side of a rotor to produce a unique pump impeller. A photograph of the pump impeller is presented in Figure 8.4.

The manufacturer's description of the pump operation, paraphrased from the operating manual sent with the pump, is presented in the following:

Model	DP-50
Power	$\frac{1}{2}$ hp
Electric	110 V; 60 hz
Speed	3450 rpm
Maximum flow	10.5 gpm
Maximum head	198 ft
Pipe diameter	1″ (Inlet and Outlet)
New weight	11 lb
Dimensions	$7″ \times 8″ \times 12″$
Features	DP-50 is a peripheral impeller pump containing numerous radial blades on either side of its impeller. The particular shape of the blades rapidly transfers radial re-circulation motion to the fluid on entry to the pump, between the impeller blades and the double channel, machined on each side of the pump body. The pump is reliable, economical and simple to use.

FIGURE 8.4
Photograph of the pump impeller.

8.5 Results and Discussion

The manufacturer's pump curve of head produced versus flow rate, plotted in Figure 8.5, was obtained from the manufacturer's "Clean Water Pump Operation Manual," which was shipped in the shipping container with the pump.

The data reduction from this study was accomplished by writing an Excel program, which is fully documented here in Figure 8.6. The pump and efficiency curves from that program are presented in Figures 8.5 and 8.6.

A comparison of the heads indicated by the manufacturer and by this study is shown in Table 8.1. At high flow rates, the manufacturer's heads are almost twice the experimentally measured heads. The best efficiency point is at 20 $\frac{L}{min}$ (5.3 gpm), and the best efficiency is 11%.

This regenerative pump is inefficient compared to pumps that are normally used for higher flows. There is evidently significant energy loss as the fluid circulates within the cavities between the impeller blades, the center rotor and the pump housing. However, for a small pump, the head developed is impressive and these pumps find many applications for producing high heads at low flows. They are used extensively in lieu of positive displacement pumps (e.g., gear-, flexible impeller-, piston-, and roller-type pumps).

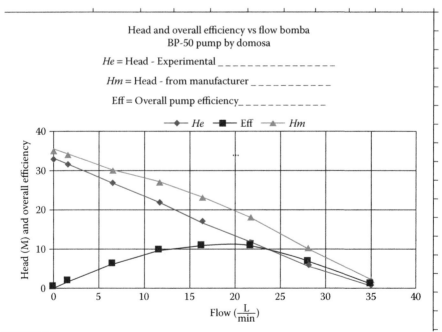

FIGURE 8.5
Results of the current study for the experimentally determined pump curve (Head vs. Flow—H_e), the pump curve from the manufacturer's operating manual (H_m), and the overall efficiency curve (Efficiency vs. flow).

	A	B	C	D	E	F	G	H	I	J	K	L	M
1	DATA REDUCTION FOR THE PUMP AND EFFICIENCY CURVES - BOMBA PUMP DP-50												
2	Pd (psi)	W (lb)	Time (s)	Watts	H (m)	KgPm	LpM	Work	Eff	LpM	He	Eff	Hm
3	1	25.2	19.65	374	0.48	35	35	2.77	0.7	35	0.5	0.7	2
4	8.3	24.7	23.94	375	5.6	28.1	28.1	25.8	6.9	28.1	5.6	6.9	10
5	17.1	23	28.91	381	11.8	21.7	21.7	41.7	11	21.7	12	11	18
6	24	22.8	37.97	405	16.6	16.4	16.4	44.4	11	16.4	17	11	23
7	31.3	25.4	59	430	21.7	11.7	11.7	41.7	9.7	11.7	22	9.7	27
8	31.3	25.8	60.12	430	21.7	11.7	11.7	41.5	9.7	11.7	22	9.7	27
9	38.6	25.1	103.25	486	26.8	6.63	6.63	29.1	6	6.63	27	6	30
10	45.5	9.4	168.5	530	31.7	1.52	1.52	7.87	1.5	1.52	32	1.5	34
11	47.1	0	999999	580	32.5	0	0	0	0	0	33	0	35
12	H = A2*(101000/14.7)/1000/9.81 - 8.5*0.0254												
13	KgPm = (B2/C2/2.2)*60												
14	LpM = (B2/C2/8.33)*60*3.785												
15	Work = (F2/60)*9.81*E2												
16	Eff = 100*H2/D2												

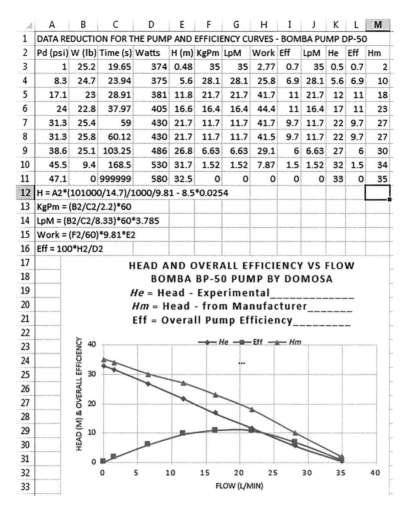

HEAD AND OVERALL EFFICIENCY VS FLOW
BOMBA BP-50 PUMP BY DOMOSA
He = Head - Experimental_____
Hm = Head - from Manufacturer_____
Eff = Overall Pump Efficiency_____

FIGURE 8.6
Documentation of the Excel data reduction program.

TABLE 8.1

Comparison of Manufacturer's and Experimental Heads

Flow $\left(\frac{L}{min}\right)$	0	10	20	30	35
Head by manufacturer (m)	36	28	19	8	0
Head by measurement (m)	33	28	13	4.5	0
% Difference (Manuf. versus Measured)	9	0	32	44	0

8.6 Conclusions and Recommendations

1. A test unit and a test program were developed that accurately deter-mine a pump curve (i.e., developed head vs. flow) and an overall efficiency curve (efficiency vs. flow) for a small regenerative turbine pump.

2. The tested pump does produce high heads at low flows. For exam-ple, the head produced at its best efficiency point, at 20 $\frac{L}{min}$, is 13 m (43 ft), which is quite high for such a small centrifugal pump.

3. However, the high heads are developed by sacrificing overall pump efficiency because the system efficiency at the best efficiency point is only 11%, which is very low compared to centrifugal pumps designed for larger flows.

4. This experiment could easily be used to determine the minimum NPSH for the pump by heating the water in the feed tank until cavi-tation occurred.

8.7 Nomenclature

Latin Letters

G	Gravitational constant on Earth, 9.81 $\frac{m}{s^2}$
h_f	Friction component in the Bernoulli balance, M
H_p	Head of fluid developed by the pump—suction to discharge, m
m	Mass flow rate of flowing fluid, $\frac{kg}{s}$
p_a	Entrance pressure of the system for a Bernoulli balance, Pa
p_b	Exit of the system for a Bernoulli balance, Pa
p_{il}	The pressure in the inlet of the pump suction line, Pa
p_{pg}	The pressure indicated by the pressure gauge, Pa
P_t	The theoretical power to pump the fluid, W
P_m	Electrical power input to the pump motor, W
T	Time of water flow into the weighing pail, s
V_a	Entrance velocity system for a Bernoulli balance, $\frac{m}{s}$
V_b	Exit velocity system for a Bernoulli balance, $\frac{m}{s}$
W	Weight of water measured in the weighing pail, kg
Z_a	Entrance elevation of the system for a Bernoulli balance, m
Z_b	Exit elevation of the system for a Bernoulli balance, m
Z_{il}	Elevation of the inlet line, taken as the datum; thus, = 0, m
Z_{pg}	Elevation of the pressure gauge relative to the datum, m

Greek Symbols

η   Pump efficiency, Dimensionless

ρ   Fluid density of the fluid, $\frac{kg}{m^3}$

References

McCabe, W.L., Smith, J.C, and Harriott, P. 2005. *Unit Operations of Chemical Engineering*, 7th ed., p. 125. New York: McGraw-Hill.

Milnes, M. 2017. *The Mathematics of Pumping Water*. Accessed August 25, 2017. p. 3. http://www.raeng.org.uk/publications/other/17-pumping-water.

Schroeder, T. 2014. *How to Read a Centrifugal Pump Curve*. June 26. Accessed August 25, 2017. https://blog.craneengineering.net/how-to-read-a-centrifugal-pump-curve.

Section II

Heat Transfer Experiments and Demonstrations

9

Thermal Conductivity and Radiative Experiments Measurement[*]

**Edgar C. Clausen, William Roy Penney, Dave C. Marrs,
Megan V. Park, Anthony M. Scalia, and Nathaniel S. Weston**

CONTENTS

[*] Reprinted from Clausen, E.C. et al., *Proceedings of the 2005 American Society for Engineering Education-Gulf Southwest Annual Conference,* Copyright 2005, with permission from ASEE.

9.1 Thermal Conductivity of Sheet and Granular Materials

9.1.1 Introduction

Thermal conductivity is a quantitative measure of the ability of a material to conduct heat according to Fourier's law. Thermal conductivity varies from a low value of 0.026 $\frac{W}{m\,°C}$ for rigid foam urethane to 2300 $\frac{W}{m\,°C}$ for diamond. In this experiment, a simple, inexpensive transient method was used for determining the thermal conductivity of low conductivity solid and granular materials. The objectives of this experiment were to

1. Determine the thermal conductivities of various sheet materials.
2. Compare the experimental thermal conductivities to literature values.

No attempts were made to include contact resistance due to air gaps and surface irregularities, which were of minor importance.

9.1.2 Experimental

9.1.2.1 Experimental Equipment List

- Three mill-finish aluminum plates (3.8 × 30.5 × 45.7 cm [$1\frac{1}{2}$ × 12 × 18 in])
- Two Omega Model HH12 thermocouple readers
- Two 3.2 mm diameter × 30.5 cm long ($\frac{1}{8}$ in diameter × 12 in long) sheathed thermocouples
- One hair dryer (Hartman Protec Model 1600)
- One insulated (1.3 cm [$\frac{1}{2}$ in] Styrofoam®) heating box (58.4 × 50.8 × 33.0 cm, or 23 × 20 × 13 in)
- Two Plexiglass® sheets (45.7 × 30.5 × 3.2 mm [18 × 12 × $\frac{1}{8}$ in])
- Two polystyrene foam sheets (45.7 × 30.5 cm × 5.7 mm [18 × 12 × 0.23 in])
- Two plywood sheets (45.7 × 30.5 cm × 11.1 mm [18 × 12 × 0.44 in])
- Insulation sheets of 1.3 cm ($\frac{1}{2}$ in)-thick Styrofoam®
- One stopwatch

9.1.2.2 Experimental Procedure

The experimental apparatus is shown in the schematics of Figures 9.1 and 9.2 and the photographs of Figures 9.3 and 9.4.

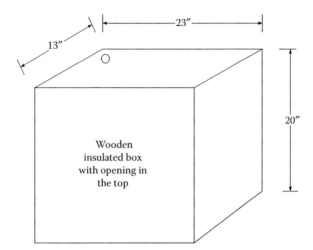

FIGURE 9.1
Insulated wooden box for heating aluminum plate.

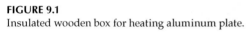

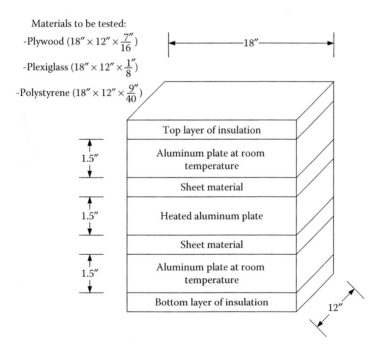

FIGURE 9.2
Schematic of test arrangement for the plates and test materials.

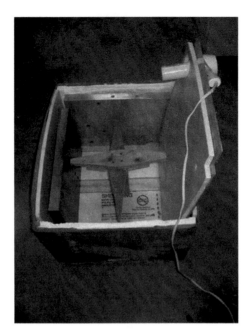

FIGURE 9.3
Photograph of wooden box used to heat aluminum plate.

FIGURE 9.4
Photograph of thermal conductivity apparatus setup.

9.1.2.2.1 Setup

1. Remove the heating box lid.
2. Place one aluminum plate in the heating box and replace the lid.
3. Insert the sheathed thermocouple into the aluminum plate.
4. Place the hair dryer in the hole on the top of the box and turn on the dryer to high speed.
5. While the plate is heating, place a layer of insulation on a room temperature table.
6. On top of the insulation, place one of the room temperature aluminum plates.
7. Place a sheet of test material (plywood, Plexiglass®, polystyrene) on top of the aluminum plate so that the edges line up with the plate.
8. When the aluminum plate in the insulated box has reached a temperature of approximately 65°C (150°F), turn off the hair dryer, remove the box cover, and remove the hot plate from the heating box using gloved hands.
9. Place the heated aluminum plate (the second aluminum plate) onto the test sheet.
10. Quickly place a second sheet of test material on top of the hot aluminum plate.
11. Place the third and final room temperature aluminum plate on top of the test material.
12. Place a layer of insulation on top of the uppermost aluminum plate.
13. Place insulation around the edges of the test sandwich.
14. Insert one sheathed thermocouple into the opening on the side of the hot center aluminum plate and another sheathed thermocouple into either the top or bottom room temperature plates.

9.1.2.2.2 Testing

1. Start the stopwatch when the sheathed thermocouples are placed in their respective aluminum plates.
2. Record the plate temperatures with time, at 1°C increments of temperature, until the center plate has reached a temperature of 30°C.

9.1.2.2.3 Safety Concerns

1. Wear safety glasses at all times.
2. Be very careful when handling the aluminum plates since they each weigh 14.4 kg (50 lb) and can break bones if dropped.
3. Always wear gloves when handling the hot aluminum plates.

9.1.3 Data Reduction

A heat balance on the center plate, with no heat generation within the plate, yields Equation 9.1:

$$-q_{out} = q_{acc} \tag{9.1}$$

The accumulation term for an individual aluminum plate is given by Equation 9.2:

$$q_{acc} = m\,Cp\,\frac{dT}{dt} \tag{9.2}$$

and the heat transferred by conduction through the test specimens from the center to the outer plates is given by Equation 9.3:

$$-q_{out} = q_{cond} = -2kA\,\frac{\Delta T}{\Delta x} \tag{9.3}$$

Substituting Equations 9.2 and 9.3 into Equation 9.1 and solving for the temporal change of temperature for the center plate gives Equation 9.4:

$$\frac{dT_c}{dt} = \frac{-2kA\,\dfrac{\Delta T}{\Delta x}}{m_c C_p} \tag{9.4}$$

Analogous equations, which are identical because of symmetry, are written for the two outer plates, giving three differential equations. The differential equations were inputted into a TK Solver® fourth-order Runga-Kutta routine for solving ordinary differential equations. Any other integration software could be used. Using a known (literature) value of k, the transient temperatures of all plates were determined and the predicted temperatures were plotted versus time. The experimental data were inputted into the TK Solver® data reduction program and were plotted on the same plot as the predicted temperatures. The thermal conductivity, k, in the model was varied until the TK model predicted the best visual fit to the experimental data.

9.1.4 Comparison of Experimental Results with Values from the Literature

The temperatures recorded from the aluminum plates for each sheet material tested are shown in Table 9.1. As an example of the procedure used to estimate the thermal conductivity, Figure 9.5 presents a transient plot of

TABLE 9.1

Experimental Temperatures Recorded from the Aluminum Plates

	Plywood			Plexiglass®			Polystyrene Foam	
Time, s	T_{cr} °C	T_{cr} °C	Time, s	T_{cr} °C	T_{cr} °C	Time, s	T_{cr} °C	T_{cr} °C
0	64.8	24	0	45.3	33	0	43.9	34
65	64		18	44		811	43	
120		25	84	43		1036		35
174	63		115		34	1692	42	
292	62		159	42		2701	41	
413		26	245	41				
424	61		298		35			
560	60		345	40				
702	59		467	39				
709		27	580		36			
854	58		627	38				
1010	57		845	37				
1050		28	1200	36				
1174	57							

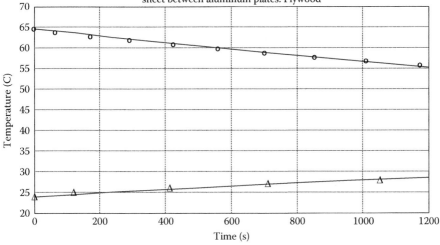

FIGURE 9.5

Transient plate temperatures with plywood ($k = 0.12\ \frac{W}{m°C}$) (Legend: Δ—outer plate; o—center plate; Lines—model predictions).

TABLE 9.2

Comparison of Calculated Experimental Thermal Conductivities
and Literature Values for Plywood, Plexiglass® and Polystyrene

	$k, \frac{W}{m°C}$		
Material	Experimental	Cengel (2003)	% Error
Plywood	0.12	0.12	0
Plexiglass®	0.24	0.19	26.3
Polystyrene Foam	0.039	0.04	2.5

temperature as a function of time for the plywood sheet. The data points
represent the experimental data from Table 9.1 and the curves show the best
fit of model using a thermal conductivity of 0.12 $\frac{W}{m°C}$.

This procedure was repeated for the Plexiglass® and polystyrene foam test
specimens.

Table 9.2 presents a comparison of the calculated experimental thermal
conductivities and the literature values, as obtained from Cengel (2003). Very
small errors (0%–2.5%) were obtained for plywood and polystyrene, while
the error for the Plexiglass® sheet (26.3%) was significantly larger but still
reasonable. These errors demonstrate that this transient technique is best
suited for materials of low thermal conductivity (<0.19 $\frac{W}{m°C}$), for which the
plate temperatures do not decrease rapidly.

9.2 Radiative Absorptivity of Metallic Surfaces

9.2.1 Introduction and Objective

Radiative absorptivity is defined as the fraction of incident irradiation absorbed
by a surface. Dewitt and Touloukian (1970) note that the radiative absorptivity

- Is influenced by the *topographical, chemical, and physical (structural)
  characteristics of the metallic surface.*
- Is one of *the most important influences on the radiative properties arising
  from surface roughness and films (oxide growth).*
- Is very *sensitive to methods of preparation, thermal history, and environmental conditions (as shown by many examples of test surfaces in the literature).*
- Is *considerably dependent upon the energy spectrum.*

Thus, one does not expect to obtain good agreement with literature values
for radiative absorptivity experiments. The objective of this investigation
was to experimentally determine the surface absorptivity of five metal surfaces relative to a matte black painted surface:

- A flat aluminum surface with a mill finish
- A cylindrical aluminum surface with a mill finish
- A cylindrical brass surface with a mill finish
- A cylindrical brass surface with an aluminum paint coating
- A cylindrical aluminum surface with a mechanically polished finish

Absorption of radiative heat from a quartz lamp by each test specimen was compared with absorption by the same test specimen painted matte black.

9.2.2 Experimental

9.2.2.1 Experimental Equipment List

The following equipment was used in the experiment:

- One Craftsman® heat lamp, with 1000 and 500 W settings
- One Omega 3.2 mm dia × 30.5 cm long ($\frac{1}{8}$ in dia × 12 in long) sheathed thermocouple
- One Omega Model HH12 thermocouple reader
- One stopwatch
- A "U"-shaped wooden support frame (Figure 9.7). The vertical 5.1 × 15.2 cm (2 × 6 in) supports are slotted to receive the 3.2 mm ($\frac{1}{8}$ in) sheathed thermocouple and the 3.2 mm ($\frac{1}{8}$ in) support rod.
- One 11.4 L (12 qt) foam cooler, partially filled with ice cubes
- One brass cylinder (1.9 cm dia × 20.6 cm long [$\frac{3}{4}$ in dia × 8 $\frac{1}{8}$ in long]), mill finish
- One brass cylinder (1.9 cm dia × 20.6 cm long [$\frac{3}{4}$ in dia × 8 $\frac{1}{8}$ in long]), painted matte black
- One brass cylinder (2.5 cm dia × 20.6 cm long [1 in dia × 8 $\frac{1}{8}$ in long]), painted matte black
- One brass cylinder (2.5 cm dia × 20.6 cm long [1 in dia × 8 $\frac{1}{8}$ in long]), painted with aluminum paint
- One aluminum cylinder (2.5 cm dia × 20.6 cm long [1 in dia × 8 $\frac{1}{8}$ in long]), mill finish
- One aluminum cylinder (2.5 cm dia x 20.6 cm long [1 in dia × 8 $\frac{1}{8}$ in long]), polished
- One aluminum rod (2.5 cm dia × 3.8 × 20.8 cm [1 × 1 $\frac{1}{2}$ in × 8 $\frac{3}{16}$ in]), with a mill-finish side and a matte black painted side
- Two 1.3 cm ($\frac{1}{2}$ in)-thick Styrofoam® insulators for the ends of the test specimens

9.2.2.2 Experimental Procedure

The experimental procedure for obtaining absorptivities involved heating each bar or rod *through* room temperature so that convective and radiative heat transfer to the surroundings were zero when the specimen passed through room temperature. Test materials of different sizes and shapes were used, along with a matte black version of the test material, to obtain the ratios of the absorptivities. The following procedure was used:

1. Cool the rods and bar (inside dry plastic baggies) in the cooler until the temperature is below 18°C.
2. Remove a test specimen from the cooler, insert a sheathed thermocouple and a support rod through 1.3 cm ($\frac{1}{2}$ in)-thick Styrofoam® insulating washers and into the center drilled holes in opposite ends of the test specimen.
3. Rest the thermocouple sheath and the support rod into the notches in the upper ends of the 5.1 × 15.2 cm (2 × 6 in) vertical support members. Align the light so that it shines directly onto the test specimen.
4. Turn on the light.
5. Start the stopwatch before the specimen reaches 18°C and record the time at each 1°C increment of temperature, to a final temperature of 30°C.

A schematic of the experimental apparatus is shown in Figure 9.6, and a photograph of the apparatus is shown in Figure 9.7.

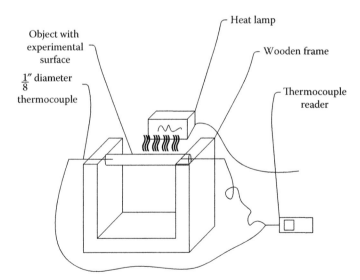

FIGURE 9.6
Schematic of experimental apparatus.

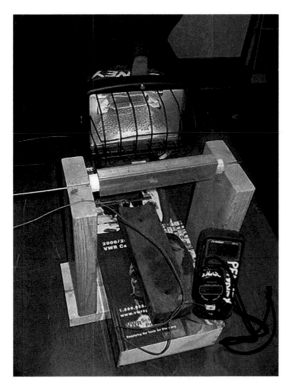

FIGURE 9.7
Photograph of experimental apparatus.

9.2.2.3 Safety Concerns

1. Wear safety glasses at all times.
2. Do not look directly into the lamp.
3. Do not place a naked hand directly in front of the lamp.
4. Do not shine the lamp directly upon any object at short range.

9.2.3 Data Reduction

To begin the data reduction, prepare a plot of temperature, T, as a function of time, t, for each of the specimens tested. Determine the slope of the T versus t curve when the specimen passes through room temperature. This determination is done at room temperature to eliminate convection and radiation to/from the surroundings.

A heat balance on the matte black specimen, at room temperature, yields Equation 9.5:

$$\alpha_{matte} q_L'' A = m_{matte} Cp, matte \left(\frac{dT}{dt} \right)_{matte} \tag{9.5}$$

where $\alpha_{matte} \approx 1$ (Cengel 2003).

Similarly, a heat balance on the test specimen yields Equation 9.6:

$$\alpha_{spec} q_L'' A = m_{spec} Cp \left(\frac{dT}{dt} \right)_{spec} \tag{9.6}$$

The ratio of Equation 9.5 to Equation 9.6 yields the absorptivity of the test specimen, shown in Equation 9.7:

$$\frac{\alpha_{matte}}{\alpha_{spec}} = \frac{m_{matte} \, C_{p,matte} \left(\dfrac{dT}{dt} \right)_{matte}}{m C_{p,spec} \left(\dfrac{dT}{dt} \right)_{spec}} \tag{9.7}$$

If the masses, areas, and specific heats are the same, Equation 9.7 reduces to Equation 9.8:

$$\frac{\alpha_{matte}}{\alpha_{spec}} = \frac{\left(\dfrac{dT}{dt} \right)_{matte}}{\left(\dfrac{dT}{dt} \right)_{spec}} \tag{9.8}$$

9.2.4 Comparison of Experimental Results with Values from the Literature

The temperatures recorded for the specimens as a function of time are shown in Table 9.3. A second-order polynomial regression was performed on each data set. The derivatives of the appropriate equations, at room temperature, gave the data needed to use Equation 9.8 to obtain $\alpha_{matte}/\alpha_{spec}$. With $\alpha_{matte} \approx 1$, α_{spec} was obtained. Table 9.4 shows a comparison of the calculated experimental absorptivities and the literature values, as obtained from Dewitt and Touloukian (1970). The errors, which ranged from 0% to 60% for all materials, are acceptable given the scope of this experiment and the difficulty in comparing data that are obtained for different light sources. This difficulty in comparing data is illustrated in the literature values for the absorptivity of polished aluminum, which differ by a factor of 3.5 (0.1–0.35, as noted by Dewitt and Touloukian [1970]). Thus, the experiment was a success in demonstrating how contrasting material surfaces alter the absorptivity of a substance.

TABLE 9.3

Specimen Temperature as a Function of Time ($T_{room} \approx 23°C$)

Painted Matte Black Aluminum Rectangular Prism (2.5 cm diameter × 3.8 × 20.8 cm [1 in diameter × $1\frac{1}{2}$ in × $8\frac{3}{16}$ in])					
t, s	*T*, °C	*t*, s	*T*, °C	*t*, s	*T*, °C
0	18	128	23	235	28
29	19	151	24	255	29
57	20	172	25	273	30
81	21	193	26	–	–
104	22	215	27	–	–

Painted Matte Black Brass Cylinder (2.5 cm diameter × 20.6 cm long [1 in diameter × $8\frac{3}{16}$ in long])					
t, s	*T*, °C	*t*, s	*T*, °C	*t*, s	*T*, °C
0	18	125	23	234	28
25	19	149	24	255	29
53	20	171	25	274	30
78	21	192	26	–	–
102	22	214	27	–	–

Brass Cylinder with Aluminum Paint (2.5 cm diameter × 20.6 cm long [1 in diameter × $8\frac{1}{8}$ in long])					
t, s	*T*, °C	*t*, s	*T*, °C	*t*, s	*T*, °C
0	18	263	23	493	28
56	19	310	24	538	29
111	20	357	25	580	30
161	21	405	26	–	–
211	22	450	27	–	–

Polished Aluminum Cylinder (2.5 cm diameter × 20.6 cm long [1 in diameter × $8\frac{1}{8}$ in long])					
t, s	*T*, °C	*t*, s	*T*, °C	*t*, s	*T*, °C
0	18	372	23	745	28
74	19	446	24	822	29
150	20	519	25	901	30
221	21	593	26	–	–
300	22	669	27	–	–

(Continued)

TABLE 9.3 (*Continued*)

Specimen Temperature as a Function of Time ($T_{room} \approx 23°C$)

Painted Matte Black Brass Cylinder (1.9 cm diameter × 20.6 cm long [$\frac{3}{4}$ in diameter × 8 $\frac{1}{8}$ in long])					
t, s	*T*, °C	*t*, s	*T*, °C	*t*, s	*T*, °C
0	18	105	23	198	28
21	19	125	24	215	29
44	20	144	25	231	30
64	21	162	26	–	–
85	22	180	27	–	–

Mill-Finish Brass Cylinder (1.9 cm diameter × 20.6 cm long [$\frac{3}{4}$ in diameter × 8 $\frac{1}{8}$ in long])					
t, s	*T*, °C	*t*, s	*T*, °C	*t*, s	*T*, °C
0	18	158	23	296	28
33	19	187	24	322	29
68	20	226	25.4	349	30
98	21	242	26	–	–
129	22	270	27	–	–

Mill-Finish Aluminum Cylinder (2.5 cm diameter × 20.6 cm long [1 in diameter × 8 $\frac{1}{8}$ in long])					
t, s	*T*, °C	*t*, s	*T*, °C	*t*, s	*T*, °C
0	18	195	23	378	28
39	19	233	24	415	29
80	20	270	25	451	30
119	21	305	26	–	–
158	22	343	27	–	–

Mill-Finish Aluminum Rectangular Prism (2.5 cm diameter × 3.8 × 20.8 cm [$1 \times 1\frac{1}{2} \times 8\frac{3}{16}$ in])					
t, s	*T*, °C	*t*, s	*T*, °C	*t*, s	*T*, °C
0	18	298	23	577	28
62	19	356	24	633	29
123	20	412	25	688	30
182	21	466	26	–	–
241	22	523	27	–	–

TABLE 9.4

Comparison of Calculated Experimental and Literature Absorptivities

Material	Absorptivity		
	Experimental	Literature	% Error
Brass, painted aluminum	0.39	0.4–0.5[a]	−3 to −28
Polished aluminum	0.22	0.1–0.35[a]	54 to −59
Mill-finish al. cylinder	0.50	0.50[b]	0
Mill-finish brass	0.66	0.6[a]	9
Mill-finish al. bar	0.37	0.50[a]	−35

[a] Cengel, Y.A., *Heat Transfer: A Practical Approach*, McGraw-Hill, New York, 2003.
[b] Dewitt, D.P. and Touloukian, Y.S., Thermophysical properties of matter, *The TPRC Data Series: Thermal Radiative Properties of Metallic Elements and Alloys*, IFI/Plenum, New York, 1970.

9.3 Conclusions

Two simple experiments were developed for obtaining (1) thermal conductivities of sheet and granular materials and (2) radiative absorptivities of metal surfaces, which help to illustrate some of the concepts taught in undergraduate heat transfer. Very small errors (0%–2.5%) in thermal conductivity were obtained for plywood and polystyrene, while the error for the Plexiglass® sheet (26.3%) was significantly larger, but still reasonable. These errors demonstrate that this transient technique is best suited for materials of low thermal conductivity ($<0.19 \frac{W}{m°C}$), where the resulting plate temperatures change relatively slowly with time.

Absorptivity errors, which ranged from 0% to 60% for all materials, are acceptable given the scope of this experiment and the inherent variability of radiative absorptivity properties with surface topographical, chemical and structural characteristics, and radiation energy spectrum. All the absorptivity experiments were successful in demonstrating how contrasting material surfaces alter the absorptivity of the surface.

9.4 Nomenclature

Latin Letters

A Heat transfer surface area, m²

C_p Specific heat of test specimen, $\frac{J}{kg\ K}$

k Thermal conductivity, $\frac{W}{m°C}$

m Mass of test specimen, kg

q_{acc} Heat rate accumulated within a control volume, W

q_{cond} Heat transfer conduction, W

q_L'' Radiative heat flux incident on test specimen surface, $\frac{W}{m^2}$

q_{out} Heat rate leaving a control volume or surface, W

t Time, s

T Temperature, °C

T_c Temperature of the center (hot) aluminum plate, °C

T_o Temperature of the outer (cold) aluminum plate(s), °C

x Linear dimension, m

Greek Symbol

α_{matte} Absorptivity of matte black specimen

α_{spec} Absorptivity of test specimen

ΔT Temperature difference between center and outside plates

References

Cengel, Y.A. 2003. *Heat Transfer: A Practical Approach.* New York: McGraw-Hill.

Dewitt, D.P. and Touloukian, Y.S. 1970. Thermophysical properties of matter, *The TPRC Data Series: Thermal Radiative Properties of Metallic Elements and Alloys.* New York: IFI/Plenum.

10

Free Convection Heat Transfer from Plates[*]

Edgar C. Clausen and William Roy Penney

CONTENTS

---

[*] Adapted from Clausen, E.C. et al., *Proceedings of the 2005 American Society for Engineering Education-Midwest Section Annual Conference*, Copyright 2005, with permission from ASEE; Clausen, E.C. and Penney, W.R., *Proceedings of the 2006 American Society for Engineering Education Annual Conference and Exposition*, Copyright 2006, with permission from ASEE.

10.1 Free Convection Heat Transfer from an Upward-Facing Horizontal Plate

10.1.1 Introduction

Free convection heat transfer is encountered in many practical applications, including heat transfer from pipes, transmission lines, baseboard heaters, and steam radiators. Correlations are available for predicting free convection heat transfer coefficients for many different geometries. One of the important geometries is the upward-facing horizontal heated surface or plate, which is the subject of this investigation. The objectives of this experiment were to

1. Determine the experimental free convection heat transfer coefficient for the top surface of a horizontal hot plate exposed to air.
2. Modify the experimental equipment to improve the experiment.
3. Compare the results with results generated from the appropriate correlation of Churchill and Chu (Cengel 2015).

10.1.2 Experimental Equipment List

The following equipment was used in carrying out this experiment:

- Hartman Pro-Tech Model 1600 hair dryer, 1600 W
- 64.1 × 55.9 × 40.6 cm (25 $\frac{1}{4}$ in × 22 in × 16 in) cardboard heating box
- 3 cm (1 $\frac{3}{16}$ in)-thick Styrofoam® insulation, lining the cardboard box
- Wooden stand to hold and elevate the aluminum plate
- 45.7 × 30.5 × 3.8 cm (18 in × 12 in × 1 $\frac{1}{2}$ in) aluminum plate, with a black painted finish
- A 10.2 cm (4 in) deep, 3.2 mm ($\frac{1}{8}$ in)-diameter hole was drilled in the center of the long side of the plate to accommodate the thermocouple, which was fully inserted into the hole
- Omega HH12 thermocouple reader
- 61 cm (24 in) of thermocouple wire: Teflon insulated, 24 gauge, type K
- Stopwatch, graduated in 0.01 s time intervals

- Additional 3 cm (1 $\frac{3}{16}$ in)-thick Styrofoam® sheet insulation
- Drop cloth curtain, for improvement of experiment

10.1.3 Experimental Procedure

The initial experimental apparatus is shown in the schematics of Figures 10.1 and 10.2 and the photographs of Figures 10.3 and 10.4.

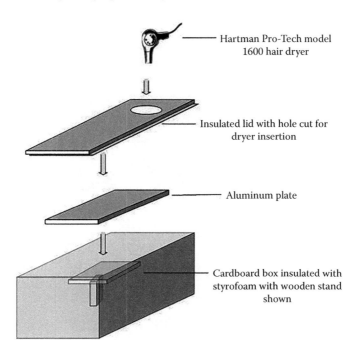

FIGURE 10.1
Insulated wooden box for heating the aluminum plate.

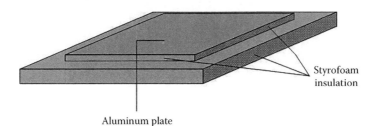

FIGURE 10.2
Experimental setup for cooling the horizontal insulated plate.

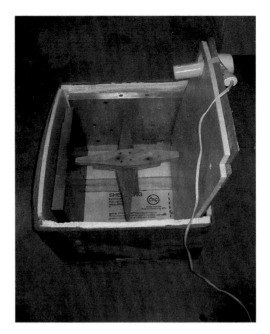

FIGURE 10.3
Photograph of wooden box used to heat the aluminum plate.

FIGURE 10.4
Photograph of apparatus for cooling the insulated horizontal plate.

10.1.3.1 Setup

1. Determine the weight of the aluminum plate and the surface area of the 30.8 × 45.7 cm (12 in × 18 in) face of the plate.
2. Make sure the air conditioning systems are off.
3. After placing the aluminum plate inside the insulated heating box, place the nozzle of the operating hair dryer into the hole in the lid, and heat the plate to ~65°C (150°F).
4. Using insulated gloves, set the aluminum plate on a sheet of Styrofoam® insulation, and wrap the two 30.5 × 3.8 cm (12 in × 1 $\frac{1}{2}$in) and two 45.7 × 3.8 cm (18 in x 1 $\frac{1}{2}$ in) faces with insulation.
5. Connect the sheathed thermocouples to the thermocouple reader and insert it into the drilled hole in the long side of the plate.

10.1.3.2 Testing

1. Start the stopwatch as soon as the temperature changes 1°C from its original temperature.
2. Record the time at each successive 1°C change in temperature.
3. Repeat the experiment as necessary.

10.1.3.3 Safety Concerns

1. Wear safety glasses at all times.
2. Be very careful when handling the aluminum plates since they each weigh 14.4 kg (50 lb) and can break bones if dropped.
3. Always wear gloves when handling the hot aluminum plates.

10.1.4 Data Reduction

A heat balance on the center plate, with no heat generation, yields Equation 10.1:

$$-q_{OUT} = q_{ACC} \tag{10.1}$$

The plate is cooled by free convection and radiation, as is shown in Equation 10.2:

$$q_{OUT} = q_{CONV} + q_{RAD} = hA_S\left(T_{SURFACE} - T_{\infty}\right) + \varepsilon\sigma A_S\left(T^4_{SURFACE} - T^4_{\infty}\right) \tag{10.2}$$

The plate accumulates heat with an inverse relationship to time as it cools back to room temperature, noted in Equation 10.3:

$$q_{ACC} = m(C_p)\frac{dT}{dt} = \rho V(C_p)\frac{dT}{dt} \tag{10.3}$$

Thus, the heat balance of Equation 10.1 yields Equation 10.4:

$$-\left(hA_S(T_{SURFACE} - T_\infty) + \varepsilon\sigma A_S(T_{SURFACE}^4 - T_\infty^4)\right) = \rho V(C_p)\frac{dT}{dt} \tag{10.4}$$

Experimental data of temperature versus time were used to determine the *best fit* experimental heat transfer coefficient by integrating Equation 10.4 numerically using a TK Solver® fourth-order Runga-Kutta integration. Other integration techniques may be used.

The heat transfer coefficient from the literature was determined using the correlation for free convection from a horizontal heated, upward-facing plate (Cengel 2003, Table 9.1, p. 542), shown in Equations 10.5 and 10.6:

$$Nu = 0.54Ra^{\frac{1}{4}} \qquad 10^4 < Ra < 10^7 \tag{10.5}$$

$$Nu = 0.15Ra^{\frac{1}{4}} \qquad 10^7 < Ra < 10^{11} \tag{10.6}$$

where the Rayleigh number is calculated as in Equation 10.7:

$$Ra = \frac{g\beta(T_{SURFACE} - T_\infty)L^3}{v^2}Pr \tag{10.7}$$

In Equation 10.7, the length of the plate is the characteristic length in free convection and, for a horizontal flat plate, $L = \frac{A_s}{P}$. Assuming that the surrounding air is an ideal gas, the volumetric expansion coefficient may be calculated in Equation 10.8:

$$\beta = \frac{1}{T} \tag{10.8}$$

Finally, h_{CORR} may be calculated from the Nusselt number as shown in Equation 10.9:

$$h_{CORR} = \frac{kNu}{L} \tag{10.9}$$

The experimental coefficient will be higher than the coefficient calculated from a literature correlation since it is impossible to remove all forced convection influences and achieve only free convection. Consequently, a factor times the theoretical coefficient was used to obtain the best fit of Equation 10.4 to the experimental data.

10.1.5 Comparison of Experimental Results with Values from the Literature

Figure 10.5 shows a plot of the experimental temperature as a function of time, as well as a curve showing a numerical integration of Equation 10.4 using the *best fit* experimental heat transfer coefficient. The emissivity (ε) of the black painted surface was assumed to be 0.98. The experimental heat transfer coefficient was $8\frac{W}{m^2K}$ at a surface temperature of 352 K, while the coefficient based on the Churchill/Chu relationship was $5.6\frac{W}{m^2K}$. Thus, a correction factor of 1.4 was needed in order to match the experimental data with the correlation. The need for this correction arises from unavoidable forced convection. It is very difficult to obtain and keep an ideal, free convection atmosphere due to existing air currents. However, isolating the apparatus in an enclosed space, turning off all air conditioning systems, and preventing any disturbances caused by movement of any kind kept these air currents to a minimum.

10.1.6 Improving the Experiment

The experimental equipment was modified to better isolate the apparatus from air disturbances. A photograph of this modification is shown in Figure 10.6, where a drop cloth curtain was used to isolate the apparatus. All other procedures were the same as described earlier, and the method of experimental analysis was also the same.

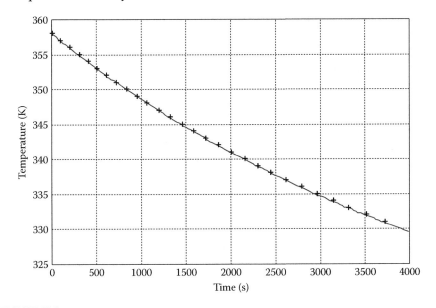

FIGURE 10.5

Temperature versus time experimental data (+) and predicted by Equation 10.4 multiplied by a factor of 1.4 ($h_{EXP} = 8\frac{W}{m^2K}$ at $T_s = 352$ K).

FIGURE 10.6
Photograph of drop cloth used to isolate the environment surrounding the aluminum plate.

When the drop cloth was added, the correction factor fell to 1.2, indicating that the addition of the drop cloth was significant in eliminating air currents.

10.2 Free Convection Heat Transfer from a Vertical Plate

10.2.1 Introduction

Another important geometry for free convection heat transfer is the vertical heated plate, the subject of this investigation. The objectives of this experiment were to

1. Determine the experimental free convection heat transfer coefficient for the vertical surfaces of a vertical hot plate exposed to air.
2. Compare these results with results generated from the appropriate correlation of Churchill and Chu (Cengel 2003, Table 9.1, p. 542).

10.2.2 Experimental Equipment List/Experimental Procedure/Safety Concerns

With the exception of a plate stand to orient the plate in a vertical fashion, the equipment, procedures and safety concerns were the same as in the previous experiment. The experimental apparatus is shown in the schematic of Figure 10.7 and the photograph of Figure 10.8.

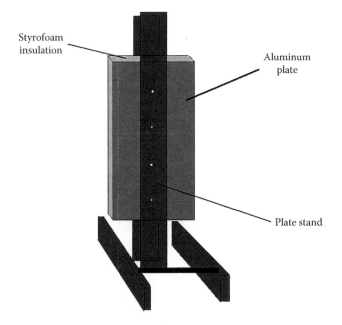

FIGURE 10.7
Schematic of vertical cooling apparatus.

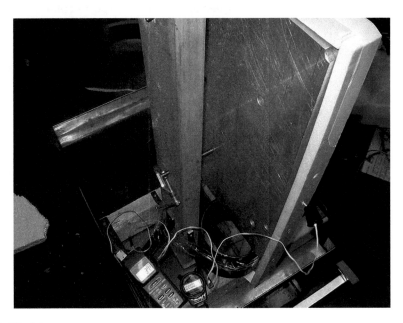

FIGURE 10.8
Photograph of vertical cooling apparatus.

10.2.3 Data Reduction

Equations 10.1 through 10.4 were used to model the transient cooling of the plate as in the previous experiment. The heat transfer coefficient from the literature was determined using the Churchill/Chu correlation, found in Equation 10.10, for free convection from a vertical plate (Cengal 2003, Table 9.1, page 542, Equation 9.21):

$$Nu = \left\{ 0.825 + \frac{0.387 Ra^{\frac{1}{6}}}{\left[1 + \left(\frac{0.492}{Pr} \right)^{\frac{9}{16}} \right]^{\frac{8}{27}}} \right\}^2 \tag{10.10}$$

10.2.4 Comparison of Experimental Results with Results Obtained from Literature Correlations

Figure 10.9 presents a plot of the experimental temperature as a function of time, as well as a curve showing a numerical integration of Equation 10.4 using the *best fit* experimental heat transfer coefficient. The experimental heat transfer coefficient was 4.7 $\frac{W}{m^2 K}$ at a surface temperature of 324.25 K, and the coefficient based on the Churchill/Chu relationship was also 4.7 $\frac{W}{m^2 K}$. Thus, the correction factor needed in order to match the experimental data with

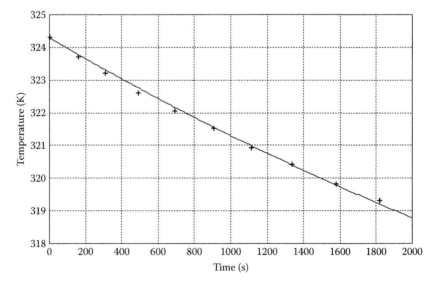

FIGURE 10.9
Temperature versus time for cooling of a vertical plate: Experimental data (+) and prediction by Equation 10.4 with a multiplication factor of 1 ($h_{EXP} = 4.7 \frac{W}{m^2 k}$, $T_s = 324.25$ K).

the correlation was 1.0. The experiment was conducted in a still storage room that had no air circulation and was completely isolated from traffic. However, the table top on which the experiment was performed retarded heat flow a bit. An emissivity (ε) of 0 was used, assuming the plate surfaces were oxidized alloy (Omega Engineering 2017). Suggestions for improving the experiment are to use polished aluminum plates or aluminum painted surfaces, and to move the plate farther from the table surface.

10.3 Conclusions

Two simple free convection heat transfer experiments were developed for determining heat transfer coefficients from an upward-facing, cooling horizontal plate and a cooling vertical plate. In determining the experimental free convection heat transfer coefficient for the top surface of a horizontal hot plate exposed to air, a correction factor of 1.4 was needed in order to match the experimental data with the correlation. The need for this correction arises from introduced forced convection. When repeating the experiment after installing curtains to minimize air circulation, the correction factor fell to 1.2.

In determining the experimental free convection heat transfer coefficient for the top surface of a vertically oriented hot plate exposed to air, a correction factor of 1.0 was needed in order to match the experimental data with the correlation; any effects of forced convection were offset by the flow-retarding effects of the table on which the vertical plate was mounted.

10.4 Nomenclature

Latin Letters

A_S	Area for convection, m^2
C_p	Specific heat of the aluminum plate or cylinder, $\frac{J}{kg\,K}$
g	Gravitational constant, $\frac{m}{s^2}$
h	Convection heat transfer coefficient, $\frac{W}{m^2 K}$
h_{CORR}	Correlated heat transfer coefficient, $\frac{W}{m^2 K}$
h_{EXP}	Experimental heat transfer coefficient, $\frac{W}{m^2 K}$
k	Fluid thermal conductivity, $\frac{W}{mK}$
L	Length of the plate or cylinder, m
m	Mass of the plate or cylinder, kg
Nu	Nusselt number

P	Perimeter of rectangular plate, m
Pr	Prandtl number of the fluid
q_{OUT}	Heat transfer out of the system, W
q_{ACC}	Heat accumulated in the system, W
q_{CONV}	Heat transfer by convection, W
q_{RAD}	Heat transfer by radiation, W
Ra	Rayleigh number of the fluid
$T\infty$	Temperature of the surroundings, K
$T_{SURFACE}$	Temperature at the surface of the plate or cylinder, K
V	Volume of the plate or cylinder, m^3

Greek Symbols

β	Volumetric expansion coefficient, K^{-1}
ε	Emissivity of the surface
μ	Dynamic viscosity of air, $\frac{Ns}{m^2}$
ν	Kinematic viscosity of air, $\frac{m^2}{s}$
ρ	Density of the aluminum plate or cylinder, $\frac{kg}{m^3}$
σ	Stefan–Boltzmann constant, $\frac{W}{m^2 K^4}$

Acknowledgment

This chapter contains results from laboratory reports prepared by a number of former University of Arkansas, Ralph E. Martin Department of Chemical Engineering, students including Cole E. Colville, Alison N. Dunn, Noor M. El Qatto, Crystal D. Hall, W. Brent Schulte, and Christopher A. von der Mehden. The authors are extremely grateful for the hard work and dedication of these University of Arkansas graduates.

References

Cengal, Y.A. 2003. *Heat Transfer: A Practical Approach*. New York: McGraw-Hill.
Cengel, Y.A. 2015. *Heat Transfer: A Practical Approach*. 5th ed. New York: McGraw-Hill.
Omega Engineering. 2017. *Emissivity of Common Materials*. Accessed August 14, 2017. https://www.omega.com/literature/transactions/volume1/emissivitya.html.

11

Forced Convection Heat Transfer by Air Flowing over the Top Surface of a Horizontal Plate[*]

Edgar C. Clausen and William Roy Penney

CONTENTS

11.1 Introduction

Forced convection heat transfer occurs when the fluid surrounding a surface is set in motion by an external means such as a fan, pump, or atmospheric disturbances. This study was concerned with forced convection

---

[*] Clausen, E.C. et al., *Proceedings of the 2005 American Society for Engineering Education-Midwest Section Annual Conference, 2005*, Copyright 2005, with permission from ASEE; Clasuen, E.C. and Penney, W.R., *Proceedings of the 2006 American Society for Engineering Education Annual Conference and Exposition*, Copyright 2006, with permission from ASEE.

heat transfer from a fluid (air) flowing parallel to a flat plate at varying velocities. The objectives of this experiment were to

1. Determine the experimental forced convection heat transfer coefficient for parallel flow over a flat plate.
2. Compare the experimental heat transfer coefficient with the coefficient calculated from the correlations presented by Cengel (2003).

11.2 Experimental Equipment List

The following equipment was used in the performance of this experiment:

- Four mill-finish aluminum plates (45.7 × 30.5 × 3.8 cm [18 × 12 × $1\frac{1}{2}$ in])
- Four 33.0 × 48.3 cm (13 × 19 in) sheets of 1.3 cm ($\frac{1}{2}$ in)-thick Styrofoam® insulation
- Thermocouple reader (Omega HH12)
- 3.2 mm dia × 30.5 cm long ($\frac{1}{8}$ in dia × 12 in long) sheathed thermocouples
- Anemometer-thermometer (Kane–May, model KM4107, serial #34095)
- 1600 W hair dryer (Hartman Protec 1600)
- Styrofoam® insulated heating box (33.0 × 50.8 × 58.4 cm [13 × 20 × 23 in])
- Stopwatch, graduated in 0.01 s time intervals
- 3-speed Black & Decker window fan, model DTS50D/B

11.3 Experimental Procedure

The schematic drawings of the experimental apparatus are presented as Figures 11.1 and 11.2 and photographs are presented as Figures 11.3 and 11.4.

11.3.1 Setup/Testing

1. Weigh each of the aluminum plates on an electronic balance. The average weight was 14.35 kg.

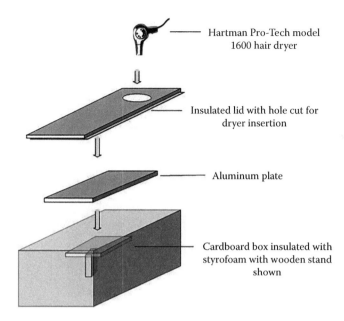

FIGURE 11.1
Insulated wooden box for heating the aluminum plates.

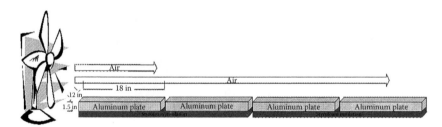

FIGURE 11.2
Location of plates for flat plate heat transfer experiment.

2. After placing two aluminum plates inside the insulated heating box, place the nozzle of the hair dryer into the hole in the lid, and heat the plates to ~65°C (150°F).

3. Place Styrofoam® insulation on a tabletop.

4. Place a heated plate in the first position on the Styrofoam® insulation, with the long (i.e., 45 cm, or 18 in) dimension in the flow direction (Figures 11.2 and 11.4).

5. Place two additional cold plates end-to-end (Figures 11.2 and 11.4), along the 45 cm (18 in) dimension, and wrap the two 30.5 × 3.8 cm (12 × 1 ½ in) and three 45.0 × 3.8 cm (18 × 1 ½ in) vertical faces with

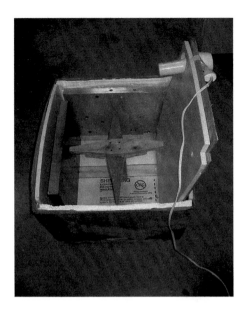

FIGURE 11.3
Photograph of insulated wood box used to heat the aluminum plates.

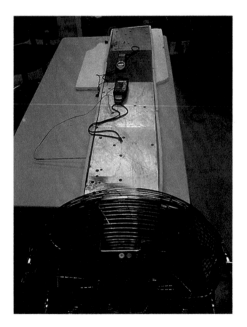

FIGURE 11.4
Photograph of experimental horizontal plate heat transfer experiment.

insulation. Leave a 1 cm space between plates to avoid conduction between the plates.

6. Connect the sheathed thermocouples to the thermocouple reader and insert them into the first plate.

7. Start the fan and choose one of three fan speeds.

8. Start the stopwatch as soon as the temperature changes 0.5°C from its original temperature.

9. Record the time at each successive 0.5°C change in temperature.

10. Use the anemometer to measure the air velocity over the plate at five different lateral positions to determine the average air velocity.

11. Repeat the above-mentioned procedure for the two other fan speed settings.

12. Remove the second heated plate from the heating box and place it in the fourth and last position from the first plate (once again, see Figures 11.2 and 11.4).

13. Repeat the above-mentioned procedures for the fourth plate.

14. Use the anemometer to measure the air velocity over the fourth plate at five different lateral positions to determine the average air velocity.

11.3.2 Safety Concerns

1. Wear safety glasses at all times.

2. Be very careful when handling the aluminum plates since they each weigh 14.4 kg (31.6 lb) and can break bones if dropped.

3. Always wear gloves when handling the hot aluminum plates.

4. Keep fingers away from the guard around the fan blades.

11.4 Data Reduction

1. A heat balance on the cooling plate, with no heat generation, yields Equation 11.1:

$$-q_{Out} = q_{Acc} \tag{11.1}$$

2. The plate is cooled by free convection and radiation, as noted in Equation 11.2:

$$q_{Out} = q_{Conv} + q_{Rad} = hA_s(T_s - T_a) + \varepsilon\sigma A_s(T_s^4 - T_a^4) \tag{11.2}$$

3. The plate accumulates heat as it cools towards room temperature, as shown in Equation 11.3:

$$q_{Acc} = -M(C_p)\frac{dT}{dt} = -\rho V(C_p)\frac{dT}{dt} \tag{11.3}$$

4. Thus, the heat balance of Equation 11.1 becomes Equation 11.4:

$$\left(hA_s(T_s - T_a) + \varepsilon\sigma A_s(T_s^4 - T_a^4)\right) = -\rho V(C_P)\frac{dT}{dt} \tag{11.4}$$

Although small, the heat balance was also corrected for the heat flow by conduction from the aluminum plate through the insulation to the table as $q_{Cond} = k_I As(T_s - T_a)/\Delta x_I$.

5. The experimental data of plate temperature versus time (in Table 11.1) were plotted using TK Solver and were curve fitted using a second-order polynomial (i.e., $T_s = a + bt + ct^2$). Other computer programs may also be used. This equation was differentiated to determine $\frac{dT}{dt} = b + 2ct$.

6. Cengel (2003) gives the following correlations in Equations 11.5 and 11.6 for *local* heat transfer coefficients for forced convection flow over a horizontal plate:

$$\mathrm{Nu}_x = \tfrac{h_x x}{k} = 0.332\,\mathrm{Re}_x^{0.5}\,\mathrm{Pr}^{\frac{1}{3}} \text{ for laminar conditions,}$$
$$\text{that is, } \mathrm{Re} < 500{,}000 \tag{11.5}$$

$$\mathrm{Nu}_x = \tfrac{h_x x}{k} = 0.0296\,\mathrm{Re}^{0.8}\,\mathrm{Pr}^{\frac{1}{3}} \text{ for turbulent conditions,}$$
$$\text{that is, } 5\times10^5 < \mathrm{Re} < 10^7 \tag{11.6}$$

The integrated *average* coefficients are given by Equations 11.7 and 11.8:

$$\mathrm{Nu} = \tfrac{hx}{k} = 0.332\,\mathrm{Re}_x^{0.5}\,\mathrm{Pr}^{\frac{1}{3}} \text{ for laminar conditions,}$$
$$\text{that is, } \mathrm{Re} < 500{,}000 \tag{11.7}$$

$$\mathrm{Nu} = \tfrac{hx}{k} = (0.037\,\mathrm{Re}^{0.8} - 871)\mathrm{Pr}^{\frac{1}{3}} \text{ turbulent conditions,}$$
$$5\times10^5 < \mathrm{Re} < 10^7 \tag{11.8}$$

TABLE 11.1

Experimental Data for Cooling of Flat Plates with Parallel Flow of Air

1st Plate		1st Plate		1st Plate		4th Plate		4th Plate		4th Plate	
V = 4.92 m/s		V = 6.00 m/s		V = 7.24 m/s		V = 3.75 m/s		V = 4.63 m/s		V = 5.54 m/s	
Time (s)	$T_s(°C)$	Time (s)	$T_s(°C)$	Time (s)	$T_s(°C)$	Time (s)	$T_s(°C)$	Time (s)	$T_s(°C)$	Time (s)	$T_s(°C)$
0	69.4	0	60.6	0	52.8	0	72.4	0	66.7	0	60.5
29	68.8	31	60.0	30	52.2	91	71.5	45	66.1	38	60.0
60	68.2	61	59.4	63	51.7	161	70.8	87	65.5	81	59.5
67	68.1	92	58.9	98	51.1	175	70.6	127	64.9	122	58.9
113	67.2	124	58.3	132	50.5	224	70.1	167	64.5	180	58.2
143	66.6	156	57.8	167	50.0	267	69.5	210	63.9	211	57.8
171	66.1	187	57.2	204	49.4	309	69.0	255	63.3	254	57.3
202	65.5	220	56.7	237	48.9	351	68.4	293	62.8	301	56.7
235	65.0	254	56.1	277	48.3	400	67.7	339	62.2	345	56.1
263	64.5	286	55.6	316	47.8	436	67.1	384	61.6	389	55.6

11.5 Comparison of Experimental Results with Values from the Literature

Table 11.1 shows the experimental data of temperature versus time for the six experiments. Figure 11.5 presents a plot of T_s versus time for the first plate at $V = 4.82 \frac{m}{s}$. All of the data were curve fitted, as shown in Figure 11.5, and the slope of all six of the individual plots was determined at the fifth data point. This slope was used in Equation 11.4 to determine the experimental heat transfer coefficient.

Table 11.2 presents the experimental and reduced data for all of the experiments. For the first plate, the ratio of h_{EXP}/h_{CORR} was 2.72, 3, and 3.29 for air velocities of 4.82, 6, and 7.24 $\frac{m}{s}$, respectively, and, for the fourth plate, the respective values were 1.71, 2.36, and 2.25 for air velocities of 3.75, 4.64, and 5.54 $\frac{m}{s}$, respectively. Thus, for the first plate the average ratio of h_{EXP}/h_{CORR} was 3 and for the fourth plate the average ratio was 2.1.

These results indicate that the experimental apparatus did not come close to producing laminar flow over the plate. This is not very surprising, considering that the fan produces significant turbulence. The fan acts like an agitator in a mechanically agitated vessel. It produces turbulence in addition to producing directed flow along the plate. In fact, it must

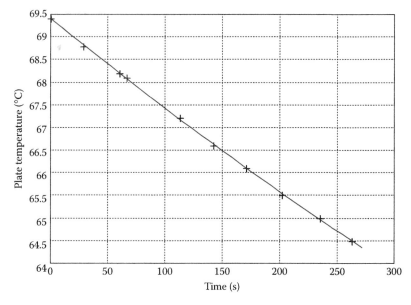

FIGURE 11.5

Temperature versus time experimental data from the first plate at an air velocity of 4.82 $\frac{m}{s}$.

TABLE 11.2

Reduced Data for All Experiments—Air Flow over Flat Plate

Plate	V	Re	Nu_x	h_{CORR}	T_s	dT/dt	q	F_{CONV}	F_{RAD}	F_{COND}	h_{EXP}
1st	4.82	1.3E5	212	12.1	67.2	−0.019	245	0.83	0.09	0.08	33
1st	6.00	1.5E5	234	13.3	58.3	−0.018	228	0.85	0.08	0.07	40
1st	7.24	1.8E5	257	14.6	50.5	−0.016	209	0.88	0.06	0.06	48
4th	3.75	3.3E5	73	9.9	70.1	−0.012	157	0.71	0.16	0.13	17
4th	4.63	4.1E5	192	11.0	64.5	−0.015	190	0.79	0.11	0.10	26
4th	5.54	4.9E5	210	12.0	58.2	−0.013	164	0.80	0.11	0.09	27

produce a great deal of turbulence for the measured coefficients to be 200%–300% higher than those that would be produced by non-turbulent laminar flow.

11.6 Improved Experiment

As is shown in Figure 11.6, a series of cardboard honeycomb diffusers were used in an attempt to minimize air turbulence. The diffusers were located just after the fan (connected to the fan outlet by a plastic *garbage bag* channel)

FIGURE 11.6
Photograph of diffuser and connection between fan and diffuser.

and immediately in front of the aluminum plates. When the air diffuser was added, the ratio was 1.8–1.9 for the first plate and ranged from 2.8 to 5.0 (average of 3.9) for the fourth plate. Thus, the diffuser only marginally affected the effects of air turbulence on temperature measurement. Perhaps the addition of a drop cloth in combination with the diffuser would have improved the ratio.

11.7 Conclusions

A simple forced convection heat transfer experiment was developed for air flowing over an upward-facing cooling horizontal plate. The experimental heat transfer coefficients were compared with correlation-predicted values from the literature. The experimental coefficients for the flat plate in parallel flow were 2.2–3.5 times higher than the literature correlation coefficients, primarily because the flow from the fans was highly turbulent and the literature correlations were for laminar conditions. When a series of cardboard honeycomb diffusers was used in an attempt to minimize air turbulence, only marginal differences were observed in the heat transfer coefficients.

11.8 Nomenclature

A_S	Heat transfer area, m^2
C_p	Specific heat, $\frac{J}{kg\,K}$
F_{CONV}	Fraction of total heat transfer by convection
F_{COND}	Fraction of total heat transfer by conduction
F_{RAD}	Fraction of total heat transfer by radiation
h	Area average convection heat transfer coefficient, $\frac{W}{m^2\,K}$
h_{CORR}	Heat transfer coefficient from literature correlations, $\frac{W}{m^2\,K}$
h_{EXP}	Heat transfer coefficient from experimental data, $\frac{W}{m^2\,K}$
h_x	Local heat transfer coefficient at length x along a flat plate, $\frac{W}{m^2\,K}$
k	Fluid thermal conductivity, $\frac{W}{mK}$
M	Mass of the plate or cylinder, kg
Nu	Area average Nusselt number, $\frac{h_x}{k}$ or $\frac{h_D}{k}$
Nu_x	Local Nusselt number at location x along flat plate, $\frac{h_x}{k}$

Pr	Prandtl number of the fluid
q_{Out}	Heat transfer from the system, W
q_{Acc}	Heat accumulated within the system, W
q_{Conv}	Heat transfer by convection, W
q_{Rad}	Heat transfer by radiation, W
Re	Reynolds number, $= \frac{VD\rho}{\mu}$ for cylinder and $\frac{Vx\rho}{\mu}$ for a flat plate
T_a	Ambient temperature of surroundings, K
T_{Film}	Film temperature $= \frac{(T_s+T_a)}{2}$, K
T_s	Surface temperature, K
V	Volume of plate or cylinder, m³
x	Length along flat plate in flow direction, m
ε	Surface emissivity
ρ	Fluid density, $\frac{kg}{m^3}$
σ	Stefan–Boltzmann constant, $\frac{W}{m^2 K^4}$

Acknowledgment

This chapter contains results from laboratory reports prepared by a number of former University of Arkansas, Ralph E. Martin Department of Chemical Engineering, students including Alison N. Dunn, Jennifer M. Gray, Jerod C. Hollingsworth, Pei-Ting Hsu, Brian K. McLelland, Patrick M. Sweeney, Thuy D. Tran, Christopher A. von der Mehden and Jin-Yuan Wang. The authors are extremely grateful for the hard work and dedication of these University of Arkansas graduates.

Reference

Cengel, Y.A. 2003. *Heat Transfer: A Practical Approach*. New York: McGraw-Hill.

12

Laminar Forced Convection Inside a Coiled Copper Tube[*]

Edgar C. Clausen, William Roy Penney, Jeffrey R. Dorman,
Daniel E. Fluornoy, Alice K. Keogh, and Lauren N. Leach

CONTENTS

12.1 Introduction

Convection is one of the primary modes of heat transport. Convection can occur as either free or forced convection in either the laminar or turbulent flow regimes. A Google® search of *free and forced convection* returns 815,000 results, indicating that convection is truly important in many aspects of technology.

The objectives of this experiment were to

1. Determine experimental heat transfer coefficients for laminar flow inside a coiled tube.

2. Compare the experimental coefficients with those predicted by the Seider–Tate equation, which is applicable for laminar tube flow.

[*] Reprinted from Foley, J.N. et al., *Proceedings of the 2015 American Society for Engineering Education Zone III Conference*, Copyright 2015, with permission from ASEE; Penney, W.R. et al., *Proceedings of the ASEE Midwest Regional Conference*, Copyright 2016, with permission from ASEE.

12.2 Experimental

12.2.1 Experimental Equipment List

The following equipment was used in performing this experiment:

- Coiled copper tubing (6.4 mm o.d. × 4.8 mm i.d. × 3.35 m long [$\frac{1}{4}$ in o.d. × $\frac{3}{16}$ in i.d. × 11 ft long]). There were six coils, each having a diameter of 17.8 cm (7 in).
- 182 cm (6 ft) of 15.9 mm i.d. ($\frac{5}{8}$ in i.d.) Tygon® tubing.
- One thermocouple reader (Omega Model HH12).
- One 1 kW immersion heater (Fisher Automerse, 115 V, 7.5A) with variable temperature control.
- One VWR thermocouple reader (Serial 221109742).
- One Arrow Engineering variable speed agitator drive (75 W [$\frac{1}{10}$ hp], 6000 rpm max).
- One 10.2 cm (4 in) diameter six-blade disk impeller operated at about 200 rpm.
- One 18.9 L (5 gal) polyethylene pail.
- One 1 L collection beaker.
- 15.1 L (4 gal) of hot water.
- 7.6 L (2 gal) of ethylene glycol (commercial automotive antifreeze).
- Two 3.8 L (1 gal) containers.
- One stopwatch.
- One TEEL Centrifugal Pump (Model 1P676A, 1630 rpm, 13 W [$\frac{1}{55}$ hp], 115 V, 0.85A).
- One Baldor variable speed motor (0.37 kW [0.5 hp], 1750 rpm).
- One Micropump positive displacement pump (Model 21056C).

12.2.2 Experimental Apparatus

Figure 12.1 presents a process flow schematic of the experimental apparatus and Figure 12.2 is a photograph of the experimental apparatus. As the photograph shows, the suction to the glycol pump was an 8.5 mm ($\frac{1}{3}$ in) i.d, silicone tube with its inlet end immersed in a 3.8 L (1 gal) feed reservoir. The discharge of the pump was connected to the copper coil, which was placed in the heated, agitated water bath. The heated glycol solution from the coil was collected in either a 3.8 L (1 gal) jug or a 1 L sample collection beaker. The water bath was contained in an 18.9 L (5 gal) pail. The water bath agitator was supported by a tripod stand with its legs outside the pail.

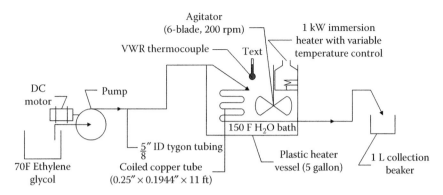

FIGURE 12.1
Schematic of experimental apparatus.

FIGURE 12.2
Photograph of experimental apparatus.

12.2.3 Experimental Procedure

1. Assemble the experimental apparatus as shown in the schematic of Figure 12.1 and the photograph of Figure 12.2.
2. Fill the reservoir with 3.8 L (1 gal) of ethylene glycol.
3. Fill the polyethylene pail with about 15 L (4 gal) of hot water.
4. Operate the agitator drive at about 200 rpm.
5. Insert the 1 kW immersion heater into the vessel, being careful to use appropriate safe supports.

6. Adjust the variable temperature knob on the heater until the water temperature reaches a steady 65°C (150°F).

7. Prime the pump by manually filling the inlet tube from the ethylene glycol tank to the pump with ethylene glycol.

8. Turn on the pump and allow the ethylene glycol to flow through the coil and into the 3.8 L (1 gal) collection container.

9. After about 1.9 L ($\frac{1}{2}$ gal) of ethylene glycol has flowed through the system, which is sufficient to reach steady state, collect a 500 mL sample of the ethylene glycol exiting the coil. Record the collection time in order to calculate the volumetric flow rate.

10. Immediately measure the feed sample temperature with the thermocouple.

11. Measure the temperature of the effluent ethylene glycol in the ethylene glycol feed tank.

12. Repeat the experiment at several different ethylene glycol flow rates.

12.2.4 Safety Concerns

1. Wear safety glasses at all times.

2. Avoid skin exposure to ethylene glycol. Should exposure occur, wash the affected area with soap and water.

12.3 Data Reduction

(NOTE: Tube wall and outside tube thermal resistances were ignored.)

1. Calculate the duty of the copper coil using Equation 12.1:

$$q_{out} = \dot{m}C_p(T_{out} - T_{in}) \tag{12.1}$$

2. Calculate the surface area of heat transfer using Equation 12.2:

$$A_s = \pi d_i l_c \tag{12.2}$$

3. Calculate the LMTD, as is shown in Equation 12.3:

$$\text{LMTD} = \left| \frac{(T_s - T_{in}) - (T_s - T_{out})}{\ln\left(\frac{(T_s - T_{in})}{(T_s - T_{out})}\right)} \right| \tag{12.3}$$

4. Calculate the experimental heat transfer coefficient, as is shown in the development of Equations 12.4:

$$q_{in} = q_{out} \tag{12.4}$$

$$q_{in} = h_e * A_s * \text{LMTD}$$

$$\therefore h_e = \left(\frac{q_{out}}{A_s * \text{LMTD}} \right)$$

5. Calculate the experimental Nusselt number, as in Equation 12.5:

$$\text{Nu}_e = \frac{h_e * d_i}{k} \tag{12.5}$$

6. Calculate the ethylene glycol velocity from Equation 12.6:

$$v = \dot{m} * \left(\frac{1}{\rho * A_c * l_c} \right) \tag{12.6}$$

7. Calculate the Reynolds number using Equation 12.7:

$$\text{Re} = \frac{d_i * v * \rho}{\mu} \tag{12.7}$$

8. Calculate the experimental Nusselt number from the Seider–Tate equation (Cengel and Ghajar 2015), shown in Equation 12.8:

$$\text{Nu}_c = 1.86 \left(\frac{\text{Re} * \text{Pr} * d_i}{l_c} \right)^{\frac{1}{3}} * \left(\frac{\mu}{\mu_s} \right)^{0.14} \tag{12.8}$$

9. Calculate the correlational heat transfer coefficient using Equation 12.9:

$$h_c = \frac{\text{Nu}_c k}{d_o} \tag{12.9}$$

12.4 Comparison of Experimental Results with Correlation

Table 12.1 presents the experimental data and Table 12.2 presents the calculated correlated and experimental values. The maximum experimental error was 24%, which is very reasonable for these experiments.

TABLE 12.1

Experimental Results

$\dot{m}, \frac{\text{lb}}{\text{hr}}$	$T_s, °F$	$T_{in}, °F$	$T_{out}, °F$	$q, \frac{\text{BTU}}{\text{hr}}$	LMTD, °F
83.5	148	69.6	126	9143	43.8
59.4	139	69.6	123	6153	36.4
93.8	126	69.6	108	6934	34.1
63.6	125	69.6	109	4812	32.3
31.3	123	69.6	114	2694	24.8
45.9	121	69.6	105	3191	29.6

TABLE 12.2

Calculated Correlated and Experimental Heat Transfer Coefficients

$v, \frac{\text{ft}}{\text{s}}$	Re	Nu_e	Nu_c	$h_e, \frac{\text{BTU}}{\text{hr ft}^2°\text{F}}$	$h_c, \frac{\text{BTU}}{\text{hr ft}^2°\text{F}}$	% Error
1.66	360	12.5	10.0	376	302	19.7
1.16	182	10.3	8.1	304	240	21.1
1.86	287	12.3	9.4	366	280	23.7
1.26	195	9.0	8.3	268	246	8.5
0.62	96	6.6	6.3	196	186	−4.9
0.91	141	6.5	7.1	194	211	−8.9

12.5 Conclusions

1. The experimental equipment and test procedure is adequate to obtain experimental data under laminar flow conditions that can be compared with correlational results.

2. The maximum deviation between correlational and experimental results is about 24% and the average deviation for 6 runs is 14.5%.

12.6 Nomenclature

Latin Letters

A_c Tube cross-sectional area, m², ft²

A_s Surface area for heat transfer, m², ft²

C_p Heat capacity of fluid, $\frac{1}{\text{kgK}}$ ($\frac{\text{BTU}}{\text{lb}°\text{F}}$) (at T_m)

d_i — Inside diameter of copper coil of tube, m (ft)

h_c — Correlational heat transfer coefficient, $\frac{1}{\text{hr m}^2\text{K}}\left(\frac{\text{BTU}}{\text{hr ft}^2\,^\circ\text{F}}\right)$

h_e — Experimental heat transfer coefficient, $\frac{1}{\text{hr m}^2\text{K}}\left(\frac{\text{BTU}}{\text{hr ft}^2\,^\circ\text{F}}\right)$

K — Thermal conductivity of fluid, $\frac{1}{\text{hr ft K}}\left(\frac{\text{BTU}}{\text{hr ft}^2\,^\circ\text{F}}\right)$ (at T_m)

l_c — Length of coil, m (ft)

LMTD — Log mean temperature difference, °C (°F)

$\dot{m}$ — Mass flow rate, $\frac{\text{kg}}{\text{hr}}\left(\frac{\text{lb}}{\text{hr}}\right)$

q_{in} — Heat transferred to fluid, $\frac{1}{\text{hr}}\left(\frac{\text{BTU}}{\text{hr}}\right)$

q_{out} — Duty of the copper coil or tube, $\frac{1}{\text{hr}}\left(\frac{\text{BTU}}{\text{hr}}\right)$

T_{in} — Temperature of the inlet stream, °C (°F)

T_m — Average temperature of T_{in} and T_{out}, °C (°F)

T_{out} — Temperature of the outlet stream, °C (°F)

T_s — Temperature of water bath (surroundings), °C (°F)

v — Velocity of fluid, $\frac{\text{m}}{\text{s}}\left(\frac{\text{ft}}{\text{s}}\right)$

Dimensionless Numbers

Nu — Nusselt number

Pr — Prandtl number (at T_m)

Re — Reynolds number

Greek Symbols

μ — Viscosity of fluid, $\frac{\text{kg}}{\text{ms}}\left(\frac{\text{lb}}{\text{ft s}}\right)$ (at T_m)

$μ_s$ — Viscosity of fluid, $\frac{\text{kg}}{\text{ms}}\left(\frac{\text{lb}}{\text{ft s}}\right)$ (at T_s)

ρ — Density of fluid, $\frac{\text{kg}}{\text{m}^3}\left(\frac{\text{lb}}{\text{ft}^3}\right)$ (at T_m)

Reference

Cengel, Y.A. and Ghajar, A.J. 2015. *Heat and Mass Transfer—Fundamentals and Applications.* p. 492. New York: McGraw-Hill.

13

An Experimental Testing of Turbulent Forced Convection Inside Tubes[*]

Edgar C. Clausen, William Roy Penney, Jeffrey R. Dorman,
Daniel E. Fluornoy, Alice K. Keogh, and Lauren N. Leach

CONTENTS

13.1 Introduction

Convection is the transfer of heat between a solid surface and a moving fluid. The fluid motion can be caused by gravity forces and density differences arising from temperature variations within the fluid—identified as *free convection*—or by motive forces such as fans or pumps—identified as *forced convection*. The turbulent flow experiments documented in this paper were of the forced convection nature because the test fluid, air, was forced through the experimental heat exchanger by the pressure of laboratory air.

---

[*] Reprinted from Foley, J.N. et al., *Proceedings of the 2015 American Society for Engineering Education Zone III Conference,* Copyright 2015, with permission from ASEE; Penney, S.L. et al., *Proceedings of the ASEE Midwest Regional Conference,* Copyright 2016 with permission from ASEE.

In this experiment, turbulent flow heat transfer coefficients were measured for the flow of air through a 6.4 mm ($\frac{1}{4}$ in) copper coiled tube. The air was heated by immersing the coiled copper tubing in an agitated hot water bath. The objectives of this experiment were to:

1. Determine experimental heat transfer coefficients for turbulent flow through a copper tube.
2. Compare the experimental coefficients with those predicted by the Dittus–Boelter correlation.

13.2 Experimental

13.2.1 Experimental Equipment List

- One 18.9 L (5 gal) polyethylene pail
- Two 6.4 mm o.d. ($\frac{1}{4}$ in o.d.) Tygon® tubes
- One King Instrument Company rotameter (0–226 $\frac{L}{min}$ [0–8 scfm] of atmospheric air)
- One thermocouple reader (Omega Model HH12)
- One 9.5 mm o.d. × 6.4 mm i.d. × 45.7 cm long ($\frac{3}{8}$ in o.d. × $\frac{1}{4}$ in i.d. × 18 in long) copper U-tube
- One Arrow Engineering variable speed agitator drive (75 W [$\frac{1}{10}$ hp], 6000 rpm max)
- One 10.2 cm (4 in) diameter six-blade disk impeller operated at about 500 rpm
- 15.1 L (4 gal) of hot water

13.2.2 Experimental Apparatus

A diagram of the experimental apparatus is shown in Figure 13.1, and a photograph of the apparatus is shown in Figure 13.2.

13.2.3 Experimental Procedure

The following experimental procedure was used to obtain the necessary data for determining the experimental heat transfer coefficients:

1. Assemble the experimental apparatus as shown in the schematic of Figure 13.1 and the photograph of Figure 13.2. NOTE: Figure 13.2 features a circular coil, which is much longer than the 45.7 cm (18 in) copper U-tube used for this experiment.

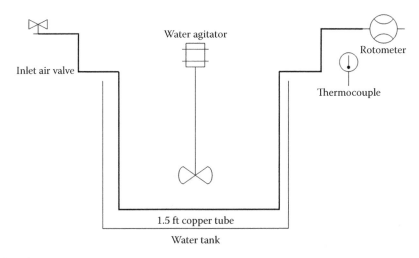

FIGURE 13.1
Schematic of experimental apparatus.

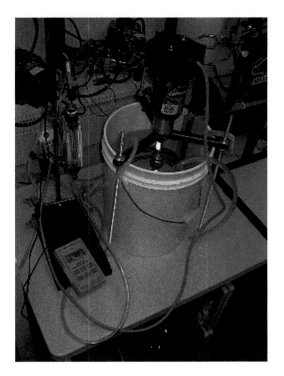

FIGURE 13.2
Photograph of the experimental apparatus.

2. Use a Tygon® tube to connect the laboratory air to the tube inlet.

3. Fill the pail with tap water at approximately 60°C (140°F) until the copper tube is completely submerged. Take care not to allow water to enter the copper tube.

4. Start the agitator and slowly increase the speed to 500 rpm. The agitator needs to operate at a relatively high speed so that the outside coefficient is very high relative to the inside coefficient.

5. Increase the air flow rate to 226 $\frac{L}{min}$ (8 cfm). Measure the inlet air temperature, allowing 1–2 min for the apparatus to reach steady state. When measuring air temperatures, be careful not to allow the thermocouple to come into contact with the Tygon® tube walls or the temperature reading will not be accurate.

6. Measure the water bath temperature.

7. Disconnect the Tygon® tube from the rotameter, measure the exit air temperature, and reconnect the tube to the rotameter.

8. Repeat steps 4–6 for flow rates in increments of 28 $\frac{L}{min}$ (1 cfm), and then duplicate the runs at 56, 112, 168, and 226 $\frac{L}{min}$ (2, 4, 6, and 8 cfm) to check experimental accuracy.

13.2.4 Safety Concerns

1. Wear safety glasses, long pants and closed-toe shoes at all times.

2. Use caution when handling hot water.

13.3 Data Reduction

(NOTE: Tube wall and outside tube thermal resistances were ignored.)

1. Calculate the duty of the copper U-tube using Equation 13.1:

$$q_{out} = \dot{m} C_p (T_{out} - T_{in}) \tag{13.1}$$

2. Calculate the U-tube surface area for heat transfer using Equation 13.2:

$$A_s = \pi d_i l_c \tag{13.2}$$

3. Calculate the LMTD, as is shown in Equation 13.3:

$$LMTD = \frac{(T_s - T_{in}) - (T_s - T_{out})}{\ln\left(\frac{(T_s - T_{in})}{(T_s - T_{out})}\right)}$$ (13.3)

4. Calculate the experimental heat transfer coefficient, as is shown in the development of Equation 13.4:

$$q_{in} = q_{out}$$ (13.4)

$$q_{in} = h_e * A_s * LMTD$$

$$\therefore h_e = \left(\frac{q_{out}}{A_s * LMTD}\right)$$

5. Calculate the experimental Nusselt number, as in Equation 13.5:

$$Nu_e = \frac{h_e * d_i}{k}$$ (13.5)

6. Calculate the air velocity from Equation 13.6:

$$v = \dot{m} * \left(\frac{1}{\rho * A_c * l_c}\right)$$ (13.6)

7. Calculate the Reynolds number using Equation 13.7:

$$Re = \frac{d_i * v * \rho}{\mu}$$ (13.7)

8. Determine the Nusselt number. The correlational Nusselt number was calculated from the Colburn equation (Cengel and Ghajar 2015), shown in Equation 13.8:

$$Nu_c = 0.023\, Re^{0.8}\, Pr^{0.4}$$ (13.8)

9. Calculate the correlational heat transfer coefficient as in Equation 13.9:

$$hc = \frac{Nu_c k}{d_o}$$ (13.9)

13.4 Comparison of Experimental Results with Correlation

Table 13.1 presents the experimental data and Table 13.2 presents the experimental and correlational heat transfer coefficients. The experimental results were very close to the correlated results, with an average error of about −5.4%. Lower volumetric flow rates (56 $\frac{L}{min}$ [2 cfm] and below) resulted in experimental heat transfer coefficients below the correlational values and higher flow rates (above 56 $\frac{L}{min}$ [2 cfm]) resulted in experimental coefficients greater than the correlational values.

13.5 Conclusions

1. A simple experiment was designed and tested to determine heat transfer coefficients for turbulent flow of air in a copper tube.

2. Accurate, consistent data were obtained.

3. The experiment could easily be conducted in a classroom by adding a portable 42 L (11 gal) air tank as an air source.

4. The experimental results agreed within a few % of correlational results obtained by use of the Colburn equation.

TABLE 13.1

Experimental Data and Selected Reduced Data

$\dot{m}, \frac{lb}{hr}$	T_s, °F	T_{in}, °F	T_{out}, °F	$q, \frac{BTU}{hr}$	LMTD, °F
34.6	132	75.2	108	273	38.1
30.3	130	75.2	107	231	36.6
26.0	129	75.2	108	205	34.9
21.7	128	75.2	107	166	34.5
17.3	127	75.2	107	132	33.4
13.0	125	75.2	107	99	31.2
8.7	125	75.2	106	64	32.0
4.3	124	75.2	106	32	30.9
8.7	123	75.2	105	62	30.5
17.3	122	75.2	104	120	30.1
26.0	121	75.2	103	174	29.8
30.3	120	75.2	102	195	29.4
34.6	119	75.2	100	207	29.7

TABLE 13.2

Calculated Correlational Results and Selected Experimental Results

$v, \frac{ft}{s}$	Re	Nu_e	Nu_c	$h_e, \frac{BTU}{hr\ ft^2 °F}$	$h_c, \frac{BTU}{hr\ ft^2 °F}$	% Error
252	37374	101	92	58.6	53.1	−9.4
220	32702	90	83	51.9	47.7	−8.2
189	28031	81	73	46.8	42.1	−9.9
157	23359	70	63	40.3	36.4	−9.7
126	18687	58	53	33.3	30.5	−8.6
94	14015	44	42	25.6	24.2	−5.3
63	9344	29	30	17.0	17.5	2.9
31	4672	15	17	8.6	10.1	17.3
63	9344	30	30	17.2	17.5	1.7
126	18687	58	53	33.7	30.5	−9.5
189	28031	82	73	47.5	42.1	−11.3
220	32702	92	83	53.2	47.7	−10.4
252	37374	101	92	58.5	53.1	−9.2

13.6 Recommendations

1. Various tube diameters could be easily tested because the tube is only 45.7 cm (18 in) long.
2. Various tube lengths could be easily tested.
3. Carbon dioxide from a laboratory cylinder could be used as a test gas.
4. Although expensive, helium would be a very effective test gas.
5. With careful data analysis, city water could be used as a test fluid. The outside tube thermal resistance would have to be determined in order to reduce the experimental data.

13.7 Nomenclature

Latin Symbols

A_c	Tube cross-sectional area, m², ft²
A_s	Surface area for heat transfer, m², ft²
C_p	Heat capacity of fluid, $\frac{1}{kgK}$ ($\frac{BTU}{lb°F}$) (at T_m)
d_i	Inside diameter of copper coil of tube, m (ft)

h_c Correlational heat transfer coefficient, $\frac{1}{hr\,m^2K}\left(\frac{BTU}{hr\,ft^2°F}\right)$

h_e Experimental heat transfer coefficient, $\frac{1}{hr\,m^2K}\left(\frac{BTU}{hr\,ft^2°F}\right)$

k Thermal conductivity of fluid, $\frac{1}{hr\,ft\,K}\left(\frac{BTU}{hr\,ft^2°F}\right)$ (at T_m)

l_c Length of coil, m (ft)

LMTD   Log mean temperature difference, °C (°F)

$\dot{m}$ Mass flow rate, $\frac{kg}{hr}\left(\frac{lb}{hr}\right)$

q_{in} Heat transferred to fluid, $\frac{1}{hr}\left(\frac{BTU}{hr}\right)$

q_{out} Duty of the copper coil or tube, $\frac{1}{hr}\left(\frac{BTU}{hr}\right)$

T_{in} Temperature of the inlet stream, °C (°F)

T_m Average temperature of T_{in} and T_{out}, °C (°F)

T_{out} Temperature of the outlet stream, °C (°F)

T_s Temperature of water bath (surroundings), °C (°F)

v Velocity of fluid, $\frac{m}{s}\left(\frac{ft}{s}\right)$

Dimensionless Numbers

Nu   Nusselt number

Pr   Prandtl number (at T_m)

Re   Reynolds number

Greek Symbols

μ Viscosity of fluid, $\frac{kg}{ms}\left(\frac{lb}{ft\,s}\right)$ (at T_m)

μ_s Viscosity of fluid, $\frac{kg}{ms}\left(\frac{lb}{ft\,s}\right)$ (at T_s)

ρ Density of fluid, $\frac{kg}{m^3}\left(\frac{lb}{ft^3}\right)$ (at T_m)

Reference

Cengel, Y.A. and Ghajar. 2015. *Heat and Mass Transfer—Fundamentals and Applications.* p. 496. New York: McGraw-Hill.

14

Forced Convection through an Annulus[*]

Edgar C. Clausen and William Roy Penney

CONTENTS

14.1 Introduction

An important geometry for forced convection heat transfer is the heating or cooling of a fluid flowing through an annulus between an outer pipe and an inner cylinder. The objectives of this experiment were to

1. Determine the experimental forced convection heat transfer coefficient for the heating of a brass rod, contained in an annulus, as air flows through the annulus.
2. Compare these results with the heat transfer coefficient from the Dittus–Boelter equation (Cengel 2003).

[*] Clausen, E.C. et al., *Proceedings of the 2005 American Society for Engineering Education-Midwest Section Annual Conference, 2005*, Copyright 2005, with permission from ASEE; Clausen, E.C. and Penney, W.R., *Proceedings of the 2006 American Society for Engineering Education Annual Conference and Exposition*, Copyright 2016, with permission from ASEE.

14.2 Experimental Equipment List

The following equipment was used in performing this experiment:

- 7.6 cm inside diameter × 183 cm long (3 in inside diameter × 72 in long) PVC tube
- 2.5 cm diameter × 108 cm long (1 in diameter × 42.5 in long) oak dowel
- 2.5 cm diameter × 20.6 cm long (1 in diameter × 8.1 in long) brass rod with a 3.2 mm diameter × 7.6 cm long ($\frac{1}{8}$ in diameter × 3 in long) center hole
- Omega HH12 thermocouple reader
- 3.2 mm diameter × 30.5 cm long ($\frac{1}{8}$ in diameter × 12 in long) sheathed thermocouple
- Hair dryer (Hartman Protec 1600)
- Stopwatch, graduated in 0.01 s time intervals
- Anemometer (Kane–May, model number KM4107)

14.3 Experimental Procedure

The experimental apparatus is shown in the schematic of Figure 14.1 and the photograph of Figure 14.2. The following procedure was used in performing the experiment:

1. Determine the weight (0.88 kg) of the brass rod and its dimensions (2.5 cm diameter × 20.6 cm long 1 in diameter × 8.1 in long).
2. Use ice to cool the rod until it is cooled below room temperature.
3. Place the wood and brass rods into the PVC tube as shown in Figure 14.1. NOTE: *The wood rod is used to provide an inside cylinder that is much longer than the brass rod, so that fully established turbulent flow exists prior to the hot air reaching the brass rod.*
4. Insert the thermocouple into the 3.2 mm ($\frac{1}{8}$ in) center hole in the brass rod.
5. Turn on the hair dryer at its highest speed, and immediately start the stopwatch.
6. Record the time for each successive 1°C change in temperature of the rod.

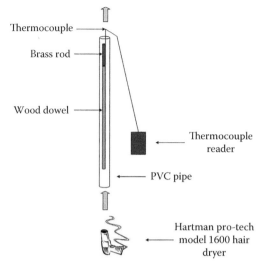

FIGURE 14.1
Schematic of annulus heating apparatus.

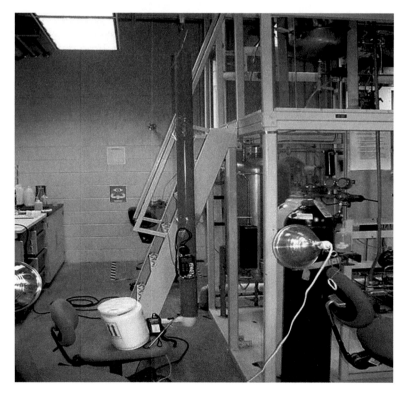

FIGURE 14.2
'Photograph' of annulus heating apparatus.

7. At a point after the air flow has reached steady state, record the velocity and ambient air temperature of the air exiting the annulus.

8. Repeat this procedure as necessary, with the same or different hair dryer speeds.

14.4 Safety Concerns

1. Wear safety glasses at all times.

2. Be on guard when the fan is used.

3. Be extra careful that the PVC outer tube is held firmly vertical against a supporting structure.

14.5 Experimental Data

Table 14.1 shows the data from the experiment, carried out at three air speeds (4.22, 2.56, and 4.43 $\frac{m}{s}$), as measured by the anemometer. The exit air temperature is shown, as well as the temperature versus time data from each experiment.

TABLE 14.1

Experimental Data of Brass Rod Temperature versus Time for Heating of the Rod with a Hair Dryer Inserted into a Pipe with the Rod in Its Center

Run #1 $V_{air} = 4.22 \frac{m}{s}$ $T_{air, out} = 62°C$		Run #2 $V_{air} = 2.56 \frac{m}{s}$ $T_{air, out} = 44.72°C$		Run #3 $V_{air} = 4.43 \frac{m}{s}$ $T_{air, out} = 64.2°C$	
Time (s)	T_r (°C)	Time (s)	T_r (°C)	Time (s)	T_r (°C)
0	12	0	10	0	13
9.6	13	47.6	11	10.9	14
27.3	14	75.9	12	24.0	15
43.6	15	103	13	46.1	16
59.1	16	130	14	62.2	17
73.6	17	160	15	77.7	18
89.1	18	187	16	92.9	19
105	19	214	17	107	20
120	20	270	18	122	21
135	21	299	19	137	22
152	22	330	20	151	23
167	23	360	21	166	24
183	24	393	22	181	25
198	25	426	23	196	26

14.6 Data Reduction

1. A heat balance on the rod with no heat generation yields Equation 14.1:

$$q_{In} - q_{Out} = q_{Acc} \tag{14.1}$$

2. The brass rod is heated by forced convection from below room temperature, through room temperature, and to above room temperature. The heat transfer coefficient is determined when the rod temperature is equal to the room temperature when heat transfer by radiation to/from the pipe walls is either 0 or negligible. Thus, Equation 14.1 becomes Equation 14.2:

$$q_{in} = hA(T_a - T_s) \tag{14.2}$$

3. The brass rod accumulates heat, as is shown in Equation 14.3:

$$q_{Acc} = m(C_p)\frac{dT_s}{dt} \tag{14.3}$$

4. Therefore, the heat balance reduces to Equation 14.4:

$$hA(T_a - T_s) = m(C_p)\frac{dT_s}{dt} \tag{14.4}$$

5. Equation 14.4 may be solved for the heat transfer coefficient, shown in Equation 14.5:

$$h = \frac{m(C_p)\dfrac{dT_s}{dt}}{A(T_a - T_s)} \tag{14.5}$$

6. The experimental data were plotted as T_s versus time and the data were curve-fitted with a second-order polynomial fit using TK Solver; that is, $T_s = a + bt + ct^2$. Any curve fitting program may be used. The slope of the curve was determined at room temperature for insertion into Equation 14.5. The plot of T_s versus t, for Run # 1, is presented in Figure 14.3. For this run, the quadratic curve fit was $T_s = 12.168 + 0.06594(t) - 6.346\text{E-}6(t^2)$, giving $\frac{dT_s}{dt} = 0.0638\frac{°C}{s}$.

7. The heat transfer coefficient from the literature was determined using the Dittus–Boelter equation (Cengel 2003) for turbulent flow through tubes with the hydraulic diameter of the annulus $(D_h = D_{pipe} - D_{rod})$ used as the characteristic length in both

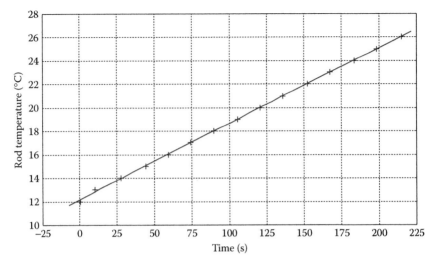

FIGURE 14.3

T_s versus time for experiment #1 with the 2.5 cm dia × 20.6 cm long (1 in dia × 8.1 in long) brass rod heated by a 62°C, 4.22 $\frac{m}{s}$ (81 $\frac{ft}{min}$) air stream in a 7.6 cm (3 in) pipe.

Re $\frac{vD\rho}{\mu}$ and Nu $\left(Nu = \left(\frac{h_{cCORR}D_h}{k}\right)\right)$. The Dittus–Boelter equation is shown in Equation 14.6:

$$Nu = 0.023\, Re^{0.8}\, Pr^{0.4} \tag{14.6}$$

8. Finally, the heat transfer coefficient from the literature correlation is calculated as shown in Equation 14.7:

$$h_{Corr} = \frac{kNu}{D_h} \tag{14.7}$$

Results from all of the annulus heat transfer experiments are summarized in Table 14.2.

TABLE 14.2

Reduced Data for All Experiments—Air Flow through an Annulus

Run	$V\left(\frac{m}{s}\right)$	Re	T_a (°C)	T_s (°C)	$\frac{dT_s}{dt}$	q	Nu	h_{CORR}	h_{EXP}	$\frac{h_{CORR}}{h_{EXP}}$
1	4.22	12,586	62.0	23	0.064	21.3	38.5	33.4	20.2	1.65
2	2.56	7,874	44.7	23	0.032	10.7	26.5	30.0	13.7	2.20
3	4.44	13,230	64.2	23	0.067	22.3	40.1	33.0	21.0	1.57

14.7 Comparison of Experimental Results with Values from the Literature

The experimental coefficients are significantly higher than the correlation-predicted coefficients. This result is not surprising, considering:

- The flow from the hair dryer is quite turbulent.
- The velocity profile from the hair dryer is not flat.
- The jet exiting the hair dryer is only 3.8 cm ($1\frac{1}{2}$ in) diameter; whereas the outside annulus pipe diameter is 7.6 cm (3 in).

The exit velocity from the hair dryer is 3.6 times the annulus velocity; this high jet velocity entering the outside annulus pipe is probably the major reason that the experimental heat transfer coefficient is so much higher than the predicted value. This entering jet would produce considerable turbulence as shear layers reduce the high jet velocity to an annulus velocity, which is only 28% of the jet velocity.

14.8 Improved Experiment

A photograph of the air diffuser, used during modification of the experiment, is shown in Figure 14.4. The diffuser was constructed from a cardboard tube, packed with soda straws, and connected to the bottom of the PVC tube in an attempt to minimize air turbulence. The procedure for obtaining the

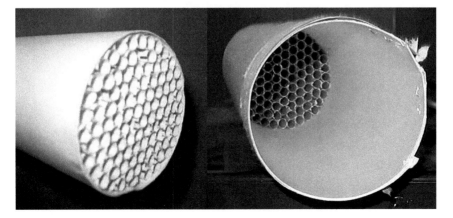

FIGURE 14.4
Two views of diffuser used in annulus heating apparatus.

experimental heat transfer coefficient was much the same as in the previous experiment. The ratio of the experimental heat transfer coefficient to the correlation heat transfer coefficient ranged from 1.6 to 2.2 for the range of air velocities, with an average of 1.8. When the air diffuser was added, the ratio held at 1.0 ($h_{EXP} = h_{CORR}$) for all air velocities, showing that this air diffuser was effective in minimizing turbulence in this system that was unaffected by outside air currents.

14.9 Conclusions

A simple forced convection heat transfer experiment was developed for hot air from a hair dryer flowing over a heated brass rod within an annulus. The experimental heat transfer coefficients were compared with literature correlation-predicted values. The experimental coefficients for the rod within an annulus were 1.6–2.2 higher than the literature correlation predictions. This finding likely results from the entering jet velocity from the hair dryer being 3.6 times the annulus velocity. This high velocity jet produces considerable turbulence as shear layers reduce the entering jet velocity to an annulus velocity that is only 28% of the jet velocity. When the air diffuser was added, the experimental and literature heat transfer coefficients were equal for all air velocities, showing that the air diffuser was effective in minimizing turbulence in the system.

14.10 Nomenclature

A_S	Heat transfer area, m^2
C_p	Specific heat, $\frac{1}{kg\,K}$
D	Cylinder diameter, m
D_h	Hydraulic diameter of the annulus, m
D_{pipe}	Pipe diameter, m
D_{rod}	Rod diameter, m
h	Area average convection heat transfer coefficient, $\frac{W}{m^2\,K}$
h_{CORR}	Heat transfer coefficient from literature correlations, $\frac{W}{m^2\,K}$
h_{EXP}	Heat transfer coefficient from experimental data, $\frac{W}{m^2\,K}$
k	Fluid thermal conductivity, $\frac{W}{m\,K}$
m	Mass of the cylinder, kg

Nu	Area average Nusselt number, $\frac{hD}{k}$
Pr	Prandtl number of the fluid
q_{In}	Heat transfer into the system, W
q_{Out}	Heat transfer from the system, W
q_{Acc}	Heat accumulated within the system, W
q_{conv}	Heat transfer by convection, W
q_{Rad}	Heat transfer by radiation, W
Re	Reynolds number $= \frac{vD\rho}{\mu}$ for cylinder
t	Time, s
T_a	Ambient temperature of surroundings, K
$T_{Air\,out}$	Temperature of air from dryer, K
T_s	Surface temperature, K
v	Fluid velocity, $\frac{m}{s}$
V	Volume of cylinder, m^3
ρ	Fluid density, $\frac{kg}{m^3}$

Acknowledgment

This chapter contains results from laboratory reports prepared by a number of former University of Arkansas, Ralph E. Martin Department of Chemical Engineering, students including Alison N. Dunn, Jennifer M. Gray, Jerod C. Hollingsworth, Pei-Ting Hsu, Brian K. McLelland, Patrick M. Sweeney, Thuy D. Tran, Christopher A. von der Mehden and Jin-Yuan Wang. The authors are extremely grateful for the hard work and dedication of these University of Arkansas graduates.

Reference

Cengel, Y.A. 2003. *Heat Transfer: A Practical Approach*. p. 496. New York: McGraw-Hill.

15

Experimental Investigation and Modeling Studies of the Transient Cooling of a Brass Rod Subjected to Forced Convection of Room Air

William Roy Penney and Edgar C. Clausen

CONTENTS

15.1 Introduction

Convection is the transport of thermal energy by a moving fluid. Forced convection occurs when the fluid movement is caused by the action of a prime mover, such as a fan. Convection is used very widely, as evidenced by a Google search of *forced convection heat transfer* that yields 1,650,000 results. Many examples of forced convection used for cooling of electronic equipment are given in Chapter 15 of Çengel and Ghajar (2010).

The purpose of this experiment was to:

1. Heat a brass rod and allow it to be cooled by forced convection of room air.
2. Measure the temperature of the rod vs. time as it cools.
3. Develop a mathematical model to predict the transient temperature of the rod as it cools.
4. Compare the results of the model prediction with the transient experimental data.

15.2 Experimental

15.2.1 Equipment List

The following equipment was used in the execution of this experiment:

- Brass rod, 207.7 g, 1.3 cm ($\frac{1}{2}$ in) diameter, 20.6 cm (8 $\frac{1}{8}$ in) long, with 3.2 mm × 6.4 cm deep ($\frac{1}{8}$ × 2.5 in deep) holes drilled in the end of the rod
- Hot wire anemometer, model number KM4107
- Fan, two-speed, Lasko Breeze Machine
- Caliper, Pittsburgh, readability to 0.0025 cm (0.001 in)
- Heat gun, Metabo, model number D-72822, 120 V AC, 1400 W
- Thermocouple reader, VWR, model number 2211109755, readability to 0.1°C
- Electronic balance, readability to 0.01 g

15.2.2 Experimental Apparatus

Photographs of the experimental apparatus are presented in Figures 15.1 and 15.2.

FIGURE 15.1
A *zoomed-out* photograph of the experimental setup.

FIGURE 15.2
A *zoomed-in* photograph of the materials used in the experiment (not pictured is the electronic balance).

15.2.3 Experimental Procedure

15.2.3.1 Equipment Setup

1. Insert a sheathed thermocouple into the drilled hole in one end of the rod.
2. Insert a 3.2 mm ($\frac{1}{8}$ in) rod into the hole in the other end of the rod.
3. Place the rod into its test position as shown in Figures 15.1 and 15.2.
4. Measure the ambient temperature using the thermocouple (22°C in this experiment).
5. Measure the diameter of the rod with a caliper and the length with a yardstick (0.5 in and 8 $\frac{1}{8}$ in, respectively, in this experiment).
6. Weigh the rod (0.2077 kg in this experiment).
7. Place the rod 91.4 cm (36 in) away from the fan.
8. Turn the fan on to setting 2.
9. Heat the rod with the heat gun by slowly moving the heat gun left to right in an even motion until the rod reaches 110°C.
10. Turn off the heat gun.

15.2.3.2 Experimental Procedure

1. Start the stopwatch when the rod cools to 100°C (Be careful not to touch the rod).
2. Observe the time in 5°C temperature increments until the rod reaches 40°C, at which time the temperature increments are changed to 2°C.
3. Stop the stopwatch when the temperature of the rod reaches 30°C.
4. Record the total time.
5. With the hot wire anemometer, measure the air velocity in front of the rod at five locations as indicated in the following:

Left Side	Left Center	Center	Right Center	Right End
3.15 $\frac{m}{s}$	3.5	3.8	3.6	2.8

The averaged air velocity was 3.37 $\frac{m}{s}$.

6. Turn off the fan.

15.2.3.3 Safety

1. Wear safety glasses, close-toed shoes, and full-length pants when in the laboratory.
2. Never eat or drink in the laboratory.
3. Sweep up long hair to the back of the head and secure with pins, net, etc.
4. Be sure to wear heat resistant gloves when handling the hot rod.
5. Do not touch the rod with bare hands.

15.3 Raw Experimental Results

The data from the experiment, as well as the data from the model prediction (described later) are shown in Table 15.1.

TABLE 15.1

Time Measurements and Predicted Times at Selected Temperatures

Temperature, °C	Experimental Time, s	Model Predicted Time, s	% Difference between Times
100	0	0	0
95	9.83	11.1	10.8
90	20.57	23.1	10.8
85	31.31	36.1	13.3
80	42.74	50.1	14.8
75	56.08	65.5	14.4
70	70.34	82.5	14.8
65	89.09	101.4	12.1
60	109.2	122.8	11.1
55	130.06	147.2	11.6
50	154.36	175.9	12.2
45	190.1	210.4	9.6
40	228.06	253.7	10.1
38	247.17	274.6	10
36	270.28	298.5	9.4
34	299.26	326.3	8.3
32	338.51	359.3	5.6
30	391.45	400.3	2.2

15.4 Model Development: Heat Transfer Coefficient at the Rod Surface

The Churchill–Bernstein equation (Çengel and Ghajar 2010) was used to calculate the heat transfer coefficient outside the rod. This equation is shown in Equation 15.1:

$$Nu = 0.3 + \frac{0.62 Re^{1/2} Pr^{1/3}}{\left[1 + \left(\frac{0.4}{Pr}\right)^{\frac{2}{3}}\right]^{1/4}} \left[1 + \left(\frac{Re}{282,000}\right)^{\frac{5}{8}}\right]^{\frac{4}{5}} \tag{15.1}$$

where the Reynolds number is described as in Equation 15.2:

$$Re = \frac{uD}{v} \tag{15.2}$$

Then, as noted in Equation 15.3:

$$h_c = \frac{Nu_f k_a}{D} \tag{15.3}$$

The surface area of the rod is shown in Equation 15.4:

$$A = \pi DL \tag{15.4}$$

A heat balance on the rod yields Equation 15.5:

$$Q_i - Q_o - Q_g = Q_a \tag{15.5}$$

The input term (Q_i) and the generation term (Q_g) are both zero for the cooling rod; thus, Equation 15.6 results:

$$-Q_o = hA(T - T_a) + \varepsilon\sigma A\left[(T + 273)^4 - (T_a + 273)^4\right] = Q_a$$

$$= mC_{pb}\left(\frac{dT}{dt}\right) = \rho V C_{pb}\left(\frac{dT}{dt}\right) \tag{15.6}$$

Equation 15.6 can be solved for $\frac{dT}{dt}$, as is shown in Equation 15.7:

$$\frac{dT}{dt} = \frac{\rho V C_p}{\left(h_c A\left(T - T_a\right) + \varepsilon \sigma A\left(\left(T + 273\right)^4 - \left(T_a + 273\right)^4\right)\right)} \tag{15.7}$$

This is the differential equation that must be solved to determine T versus t from the model. A simple Euler integration is adequate to solve this equation. For this rod, which cooled from 100°C to 30°C, a time step of 4 s is very reasonable to achieve an accurate numerical integration. The 4 s can easily be checked by changing the integration time step to 2 s and comparing the results of the 4 s run with the results from the 2 s run.

15.5 Reduced Results

In addition to showing a comparison of experimental and model results in Table 15.1, the experimental results are compared to the predicted results in Figure 15.3.

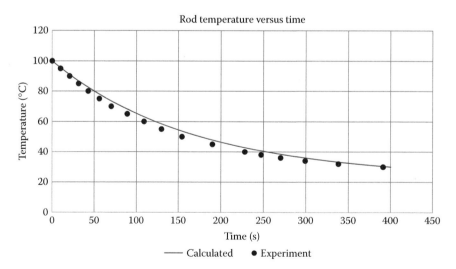

FIGURE 15.3
A graphical plot comparing the experimental temperature profile with the model-calculated temperature profile.

15.6 Discussion of Results

As evidenced by Figure 15.3, the mathematical model predicted the experimental data very well. This indicates the measurements of rod temperature and times corresponding to those temperatures were reasonably accurate. A literature value of 0.88 was used as the emissivity of the flat black painted rod.

15.7 Conclusions

1. The calculated density of the brass rod matched well with the literature density of the brass.
2. The temperature measurements and the air velocity measurements were sufficiently accurate to develop a model that predicted well the experimental transient temperature distribution.
3. The model-predicted cooling time, from 100°C to 30°C, of 400 s agreed well with the experimental cooling time of 391 s, which is only a 2.25% difference. For all 16 data points, the difference between model-predicted times and experimental times was 10%.

15.8 Recommendations

1. A polished and silvered rod could be used to reduce the radiation heat transfer.
2. The rod surface heat transfer coefficient could be measured experimentally

15.9 Nomenclature

Latin Symbols

A	Surface area of rod, m^2
$\frac{dT}{dt}$	Derivative of rod temperature versus time, $\frac{K}{s}$
D	Diameter of rod, m
C_p	Specific heat of cartridge brass, $\frac{J}{kg\,K}$

h	Convection heat transfer coefficient, $\frac{W}{m^2 K}$
k_a	Thermal conductivity of air, $\frac{W}{m\,K}$
k_b	Thermal conductivity of brass, $\frac{W}{m\,K}$
L	Length of rod, m
m	Mass of rod, kg
Nu_f	Nusselt number for forced convection
Pr	Prandtl number for air, dimensionless
$Q_i,\ Q_o,\ Q_g,\ Q_a$	In, out, generation and accumulation terms in the heat balance equation, W
Re	Reynolds number
t	Time, s
T_a	Temperature of air, C
T	Temperature of rod, C
u	Velocity of air, $\frac{m}{s}$
V	Volume of rod with hole, m^3

Greek Symbols

σ	Stefan-Boltzman constant $\frac{W}{m^2\,K^4}$
ρ	Density of cartridge brass, $\frac{kg}{m^3}$
ε	Emissivity of flat black paint brass rod, dimensionless
v	Kinematic viscosity, $\frac{m^2}{s}$

Acknowledgment

This chapter contains results from laboratory reports prepared by former University of Arkansas, Ralph E. Martin Department of Chemical Engineering, students including Richard D. Clark and Lauren N. Rogers. The authors are extremely grateful for the hard work and dedication of these University of Arkansas graduates.

References

Çengel, Y.A. and Ghajar, A.J. 2010. *Heat and Mass Transfer—Fundamentals and Applications* 5th ed., New York: McGraw Hill.

16

Solar Flux and Absorptivity Measurements[*]

William Roy Penney, Kendal J. Brown, Joel D. Vincent, and Edgar C. Clausen

CONTENTS

16.1 Introduction

All matter emits a characteristic amount of electromagnetic radiation. When this radiation is absorbed by another object, the thermal state of the object is changed and radiation heat transfer takes place. The radiation heat transfer between two objects is proportional to the difference in their temperatures raised to the fourth power, given by the Stefan–Boltzmann law (Çengel 2007), as is shown in Equation 16.1:

$$Q = \varepsilon \sigma A (T_1^4 - T_a^4) \tag{16.1}$$

When an object is exposed to a radiation flux, such as solar radiation, only a certain fraction (i.e., α) of the incident radiation is absorbed by the surface. Thus, as is noted in Equation 16.2,

$$Q_{absorbed} = \alpha Q_{incident} \tag{16.2}$$

In this experiment, Equation 16.2 was used to determine the incident solar flux by experimentally measuring the solar heat absorbed by a black painted

---

[*] Reprinted from Penney, W.R. et al., *Proceedings of the 2008 American Society for Engineering Education Midwest Section Annual Conference,* Copyright 2008, with permission from ASEE.

aluminum plate of known solar absorptivity. A thermocouple was attached to the backside of the aluminum plate to measure the transient plate temperature as it was exposed to solar radiation. To begin the experiment, the plate was first cooled to below ambient temperature and then placed inside a glass picture frame, backed with insulation. The frame was oriented toward the sun at a selected angle of incidence, and the temperature was measured as a function of time. To determine the solar flux from the experimental data of plate temperature versus time, a transient model was developed to predict the plate transient temperature. Within this model, the inputted solar flux was changed until the theoretical temperature profile closely fit the experimental data. The experiment was repeated for three angles of incidence.

The experimental solar flux was used to calculate the solar absorptivity of the mill-finish side of the aluminum plate. The temperature profile of the unpainted plate, as it was exposed to the sun, was determined in the same fashion as described above. The transient model was again used, but the absorptivity was now determined by iteration, using the experimental solar flux.

16.2 Experimental Equipment and Supplies List

The following supplies and equipment were used to perform the experiment. Of course, the size of the plate and corresponding picture frame are not critical to the experiment, and the size given below reflects what was available at the time of the experiment.

- 26 × 34 cm picture frame with a glass plate, but no backing
- Dow Styrofoam® insulation, 5.1 cm (2 in) thick
- Packing tape
- 26 × 34 cm aluminum plate, 3.2 mm ($\frac{1}{8}$ in) thick
- Non-reflective black spray paint
- Stopwatch
- Thermocouple and reader
- Watertight bag of ice

16.3 Experimental Procedure

A schematic drawing of the experimental apparatus is presented in Figure 16.1, and photographs of the experimental apparatus for both parts of the experiment are presented in Figures 16.2 and 16.3.

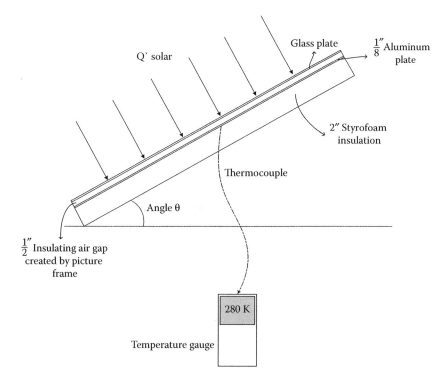

Q` solar

Glass plate

$\frac{1}{8}''$ Aluminum plate

2" Styrofoam insulation

Thermocouple

Angle θ

$\frac{1}{2}''$ Insulating air gap created by picture frame

280 K

Temperature gauge

FIGURE 16.1
Schematic of experimental apparatus.

FIGURE 16.2
Experimental setup with black side of plate facing forward.

FIGURE 16.3
Experimental setup showing unpainted side with thermocouple.

16.3.1 Setup/Testing

The experimental procedure for determining the solar flux as a function of the angle of incidence is as follows:

1. Spray paint one side of the aluminum plate with the black non-reflective paint, making sure that the entire plate is covered evenly.

2. Once the paint has dried, tape the thermocouple to the center of the unpainted side of the plate.

3. Securely attach the picture frame to the insulation using the packing tape in a manner that allows the aluminum plate to rest inside the picture frame with its back directly against the insulation.

4. Place the bag of ice on top of the black side of the plate until the temperature of the plate falls below 18°C (65°F).

5. Dry the plate and place it inside the picture frame, with the painted side facing the sun.

6. Place the glass plate in the picture frame above the aluminum plate.

7. Orient the entire apparatus normal to the sunlight by arranging it so that a rod placed normal to the surface of the plate does not cast a shadow.

8. Beginning at 18°C (65°F), record the time at roughly 2°C (1°F) increments as the plate is heated by the solar flux until the plate reaches 29°C (85°F).

9. Remove the plate from the frame and, once again, place the ice bag on the plate and leave it there until the plate reaches 18°C (65°F).

10. Repeat the experiment for other angles of incidence.

To determine the solar absorptivity of the unpainted side of the plate, repeat the experiment with the unpainted side of the plate facing the sun, and with the thermocouple attached to the painted side.

16.4 Experimental Results

Time/temperature data for the black plate at different angles of incidence from the sun are shown in Table 16.1. Similar data for the unpainted aluminum plate, placed normal to the sun, are shown in Table 16.2. As expected, the

TABLE 16.1

Experimental Time/Temperature Data for the Black Plate at Different Angles of Incidence, θ

	Time, s		
Temperature, °F	θ = 0°	θ = 45°	θ = 60°
65	0	0	0
66	6.5	4.9	6.0
67	11.3	10.3	11.9
68	15.8	14.9	18.3
69	19.3	19.5	24.6
70	21.8	24.9	31.0
71	24.0	30.0	37.3
72	27.9	35.4	44.2
73	31.5	40.4	50.5
74	36.0	45.9	56.9
75	40.3	51.8	63.7
76	44.2	57.2	70.1
77	48.8	63.2	77.4
78	53.8	68.7	84.3
79	57.6	74.6	91.0
80	62.7	80.5	99.2
81	67.0	86.9	106.0
82	71.5	92.8	113.3
83	76.6	98.8	120.7
84	81.1	105.6	128.4
85	86.1	111.6	135.3

TABLE 16.2

Experimental Time/Temperature Data for the
Aluminum Plate, Placed Normal to the Sun
($\theta = 0°$)

Temperature, °F	Time, sec
61	0
62	11.6
63	24.8
64	37.6
65	51.8
66	64.0
67	78.5
68	93.0
69	106.3
70	121.3
71	136.3
72	151.3
73	166.8
74	182.3
75	198.3
76	212.3
77	226.4
78	240.0
79	256.5
80	271.4
81	284.6

time for the black plate to heat to a final temperature of 29°C (85°F) increased
with the angle of incidence. In addition, the black plate heated much faster
than the aluminum (unpainted) plate, when both sides were oriented normal
to the sun.

16.5 Simulation and Data Reduction

To calculate the solar flux entering the aluminum plate, a heat balance was
first performed over the plate. This overall equation is shown in Figure 16.3:

$$Q_i - Q_o + Q_g = Q_a \tag{16.3}$$

With no heat generation by the plate, Equation 16.3 reduces to Equation 16.4:

$$Q_i - Q_o = Q_a \tag{16.4}$$

The heat accumulated in the plate is given by Equation 16.5:

$$Q_a = m_p C_{p,p} \frac{dT_p}{dt} = \rho_p A t_p C_{p,p} \frac{dT_p}{dt} \tag{16.5}$$

and the net solar heat entering the plate is given by Equation 16.6:

$$Q_{i,s} = \alpha A Q_S'' \tag{16.6}$$

When the plate temperature is equal to the ambient temperature, the only heat leaving the plate is by radiation to the day sky and by conduction into the Styrofoam® insulation, shown by Equation 16.7:

$$Q_o = \varepsilon \sigma A(T_p^4 - T_{sky}^4) - kA \frac{dT_{i,x=0}}{dx} \tag{16.7}$$

Substituting Equations 16.5 through 16.7 into Equation 16.4 and isolating $\frac{dt_p}{dt}$ yields Equation 16.8:

$$\frac{dT_p}{dt} = \frac{1}{\rho_p C_{p,p} A t_p} \left(\alpha A Q_S'' - \varepsilon \sigma A(T_p^4 - T_{sky}^4) - kA \frac{dT_{i,x=0}}{dx} \right) \tag{16.8}$$

The term $\frac{dt_{i,x=0}}{dx}$ in Equation 16.8 is determined by performing a transient nodal analysis for the transient temperature distribution in the insulation. The half-node convention for surface nodes was used, which means that the insulation surface node is placed at the surface, resulting in the first node temperature being equal to the plate temperature. The plate was assumed to have a uniform temperature because of the high thermal conductivity of aluminum.

The insulation internal nodes only have heat accumulation and conduction terms. The nodal accumulation term is shown in Equation 16.9:

$$Q_{a,i} = \rho_i A \Delta x_i C_{p,I} \frac{dT_i}{dt} \tag{16.9}$$

Conduction occurs to node i from nodes $i-1$ and $i+1$. The conduction term from both adjacent nodes is shown in Equation 16.10:

$$Q_{c,i} = k_i A \frac{T_{1-1} + T_{i+1} + 2T_i}{\Delta x_i} \tag{16.10}$$

Substituting Equations 16.9 and 16.10 into Equation 16.4 and isolating $\frac{dT_i}{dt}$ yields Equation 16.11:

$$\frac{dT_i}{dt} = \frac{k_i}{\rho_i \Delta x_i^2 C_{p,i}}(T_{i-1} + T_{i+1} - 2T_i) \tag{16.11}$$

The number of nodes and the nodal thickness is selected such that the temperature within the insulation is constant with depth below the last node; thus, there is essentially no heat transferred by conduction from the last node into the remaining depth of the insulation. Without conduction to the *i*+1 node, Equation 16.11 becomes Equation 16.12:

$$\frac{dT_i}{dt} = \frac{k_i}{\rho_i \Delta x_i^2 C_{p,i}}(T_{I-1} - T_I) \tag{16.12}$$

Equations 16.11 and 16.12 were solved using the ODE STIFFER integration routine in TK Solver®. Other programs may also be used. Q_S'' was found by adjusting its value until the calculated values of T_p versus t match the recorded experimental data. Figure 16.4 presents a comparison of the model-predicted plate temperature profile with the measured profile for $\theta = 0$.

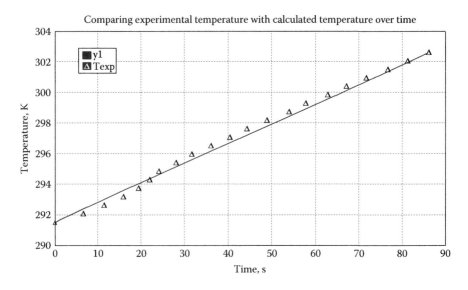

FIGURE 16.4
Experimental temperature measurements (Δ) and predicted temperatures (—) with time for the black (painted) surface placed normal to the sun ($\theta = 0°$).

In the case of the sun's rays being normal to the plate, the measured $Q''_{S,k}$ is known. Therefore, the results of the experimental $Q''_{S,e}$ can be compared with the known $Q''_{S,k}$ using Equation 16.13:

$$\% \ error = \frac{Q''_{S,e} - Q''_{S,k}}{Q''_{S,k}} \qquad (16.13)$$

For the cases where the solar fluxes are incident at 45° and 60°, Equation 16.14 is used to determine the actual solar flux:

$$Q''_{S,k,\theta} = Q''_{S,k,0} \cos \theta \qquad (16.14)$$

The above procedure was also used to analyze the data for solar flux incidence at angles of 45° and 60°. The calculated solar flux for the sun's rays normal to the plate is then used to calculate the solar absorptivity (θ) of aluminum. The absorptivity was found by varying α until the calculated variation of T_p versus t matched the recorded experimental data.

16.6 Results and Discussion

Figure 16.4 shows a plot of plate temperature versus time for the black (painted) surface placed normal to the sun ($\theta = 0°$). Both the experimental data from Table 16.1 and the predicted temperatures from the model predictions are presented in Figure 16.4. The solar flux that best fitted the data presented in Figure 16.4 was 750 $\frac{W}{m^2}$. Similar plots were obtained for the black surface placed at angles of 45° and 60°, and for the unpainted aluminum surface placed normal to the sun.

Table 16.3 presents a comparison of the experimentally determined solar fluxes and the solar fluxes measured by the National Weather Service in Westville, OK (Personal communication 2008). The National Weather Service in Westville, OK is located 37 km (23 mi) west of Fayetteville, AR (where

TABLE 16.3

Comparison of Experimental Solar Fluxes with Fluxes Generated by the National Weather Service (NWS) in Westville, OK at Different Angles of Incidence

Angle of Incidence	Solar Flux, $\frac{W}{m^2}$		% Error
	Experimental	NWS	
0°	750	750	0
45°	525	530	1
60°	390	375	4

TABLE 16.4

Comparison of Milled Aluminum Solar Absorptivities,
Determined Experimentally and from the Literature

Absorptivity	
Experimental	From EPA (2006)
0.13	0.10–0.15

the experiment was performed) and is at a similar elevation. The National Weather Service data were recorded at 11:00 a.m. on April 7, 2008, essentially the same time that the experimental data were taken. The experimental solar fluxes matched the fluxes generated by the National Weather Service with a maximum error of 4%, quite good for an experiment of this type. Table 16.4 shows a comparison of the experimentally determined solar absorptivity of the unpainted milled aluminum plate and the absorptivity from the EPA (2006). As is noted, the experimental absorptivity of 0.13 falls within the accepted range of 0.10–0.15 for milled aluminum, as found by the EPA.

16.7 Conclusions

Despite problems in accurately measuring the angle of incidence and in obtaining solar flux information for the exact site of the experimental measurements, the experimental solar fluxes agreed very well with the fluxes obtained by the National Weather Service in Westville, OK. The solar fluxes were found to decrease with increasing angle of incidence, from 750 to 390 $\frac{W}{m^2}$, as the angle of incidence was increased from $0°$ to $60°$. In measuring the solar absorptivity of mill-finish aluminum, the experimental value of 0.13 was well within the range presented by the EPA (2006).

16.8 Nomenclature

A	Area for heat transfer, m^2
C_p	Specific heat, $\frac{J}{kgK}$
k	Thermal conductivity, $\frac{W}{mK}$
m	Mass, kg
Q	Radiation heat transfer between two objects, W
Q_a	Heat accumulation, W

$Q_{absorbed}$	Radiation absorbed by a surface, W
Q_g	Heat generation, W
Q_i	Heat transferred in, W
$Q_{incident}$	Incident radiation, W
Q_o	Heat transferred out, W
Q_S''	Solar flux, $\frac{W}{m^2}$
T	Temperature, °C or K
T_{sky}	Effective sky temperature, 252 K
t	Time, s
Δx_i	Insulation nodal thickness, m

Greek Characters

α	Radiant (solar) absorptivity
ε	Emissivity
θ	Angle of incidence, °
ρ	Density, $\frac{kg}{m^3}$
σ	Stefan–Boltzmann constant, $5.67 \times 10^{-8} \frac{W}{m^2 K}$

Subscripts

i	Nodal counter, insulation, and input term in the overall heat balance
I	Total number of nodes used for the insulation nodal analysis
p	Plate

References

Çengel, Y.A. 2007. *Heat Transfer: A Practical Approach*, 3rd ed. New York: McGraw-Hill.
Environmental Protection Agency. 2006. May 30. Accessed June 24, 2008. http://www.epa.gov/ttn/chief/old/ap42/ch07/s01/draft/d7s01_table7_1_6.pdf.

17

A Transient Experiment to Determine the Heat Transfer Characteristics of a 100 W Incandescent Light Bulb, Operating at 48 W*

Lauren Cole, Lindsay R. Hoggatt, Jamie A. Sterrenberg, David R. Suttmiller, William Roy Penney, and Edgar C. Clausen

CONTENTS

17.1 Introduction

Heat transfer from a simple light bulb is pertinent in our everyday lives. Every building, every store, and every vehicle has light bulbs. Standing on a stage with full lighting overhead, or simply hovering a hand over a lit bulb, demonstrates the heat transfer from a light bulb. Incandescent light bulbs

* Reprinted from Cole, L. et al., *Proceedings of the 2012 American Society for Engineering Education Midwest Section Annual Conferece*, Copyright 2012, with permission from ASEE.

only emit about 9% of their radiation in the visible spectrum; the remainder is wasted. A significant portion of the wasted energy is absorbed by the glass envelope surrounding the filament. This absorbed energy is then transferred by convection and radiation to the environment.

The experiment described in this paper measures parameters associated with this heat transfer to determine: (1) the fraction of the filament radiation absorbed by the glass bulb, (2) the combined heat transfer coefficient for the bulb, (3) the natural convection heat transfer coefficient of the bulb, and (4) the forced convection coefficient for the bulb. This experiment is ideal for use as a laboratory experiment and/or as a classroom demonstration because of its simplicity and coverage of a number of heat transfer principles.

The incandescent light bulb is a marvel of modern technology. Riveros and Oliva (2006) present an excellent explanation of the key parts of a light bulb:

> The three main components of an incandescent lamp are: the metallic filament (usually tungsten) with a high melting point; the bulb envelope made of glass, which is empty or partially filled with an inert gas, to prevent oxidization of the element; and the base of the lamp, which includes the two separate electrodes (metallic threaded base and the eyelet) and the glass tube (which seals the lead-in wires). The pressure of the gas reduces the evaporation rate but increases the convection losses, so that lamps below 25 W are in a low vacuum. To increase the visible light production, tungsten filaments need to be heated to high temperatures (between 2700 K and 2900 K), producing a small change in colour from yellow to white for illumination. However, infrared radiation is always produced, reducing the efficiency of the visible light emission.

The glass bulb for a 100 W bulb weighs about 13 g, is 6 cm in diameter, and is about 0.4 mm thick. The glass is almost transparent to radiation in the visible spectrum, but is increasingly opaque to radiation above a 2.5 μm wavelength. A tungsten element operating at 2800 K emits a significant fraction of its radiation at wavelengths exceeding 2.5 μm. An excellent video about the complex spiral within a spiral tungsten element is available on Mr. Barlow's Blog (Hammack 2012). The glass envelope is filled to about 0.7 atm with argon/nitrogen gas to prevent evaporation of the tungsten filament. The filament operates at about 2800 K and only produces about 9% of its light in the visible spectrum. Much of the radiation from the filament is in the infrared region, where the glass is practically opaque (Çengel and Ghajar 2011).

17.2 Equipment and Procedures

The following paragraphs describe the equipment and materials, experimental procedures, and safety considerations in performing the experiment.

17.2.1 Equipment List

The equipment and supplies used in the experiment were as follows:

- 100 W, 115 V Sylvania Double Life Clear glass tungsten filament light bulbs, 2
- Omega Precision Fine Wire thermocouple, 0.008 mm (0.003 in) diameter
- Omega HH12 thermocouple reader
- Stopwatch
- Porcelain ceramic light socket
- EXTECH, Model DW-6060 wattmeter
- Digital caliper
- Laboratory ring stand
- Mettler Toledo AB104-S analytical balance
- STACO Energy Products, 120 V variable autotransformer
- Transparent tape

17.2.2 Experimental Apparatus

The experimental apparatus (Figure 17.1) consisted of a light bulb, in its socket, attached horizontally to a standard laboratory ring stand. A thermocouple was taped to the bulb as shown in Figure 17.1. A variable voltage

FIGURE 17.1
Photograph of experimental apparatus.

transformer was used to reduce the power drawn from a standard room out-let from 100 W to 48 W, as measured by the wattmeter. The reduced power was used to slow the rate at which the bulb heated as it was lit.

An identical second light bulb had its base removed for weighing the glass envelope. Figure 17.2 is a photograph of this bulb, sitting on the analytical balance. The mass of the bulb was 13.24 g. Figure 17.3 shows the caliper used to measure the bulb diameter of 6 cm.

FIGURE 17.2
Cut-away bulb being weighed.

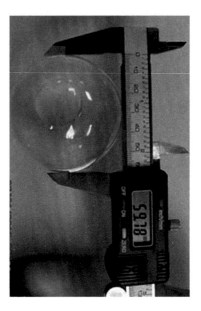

FIGURE 17.3
Caliper measuring bulb diameter.

17.2.3 Experimental Setup and Procedure

The following procedure was used in assembling the apparatus and conducting the experiment:

- A glassblower removed the metal cap and the diameter and mass of a bulb was measured.
- The light bulb was attached to the ring stand using a laboratory clamp.
- The thermocouple was taped to the bulb using transparent tape.
- The voltage toggle switch was set to "off" and the voltage control knob was rotated to 30%.
- The ambient temperature was recorded.
- The voltage toggle switch was moved to 120 V, giving a power output of 48 W.
- The time was recorded for each 3°C increase in bulb temperature, until the temperature reached 60°C; from 60°C to 80°C, the time was recorded for each 2°C rise; and, above 80°C, the time was for each 1°C rise until the bulb temperature reached steady state, as determined by nearly constant temperature with time.
- The experiment was repeated.

17.2.4 Safety

- Safety glasses, closed-toed shoes, and long pants were worn during the experiment.
- Caution was exercised to avoid touching the hot bulb.
- The bulb was allowed to become cool to touching before removing it from the ring stand.
- Care was taken in handling the cutaway light bulb envelope to avoid contact with the sharp cutaway edges.

17.3 Experimental Results

Table 17.1 presents the experimental results as incremental times between temperatures. The two incremental times were averaged and an elapsed time was computed and included in Table 17.1.

TABLE 17.1

Raw Experimental Data

Temp (°C)	Time Run 1 (s)	Time Run 2 (s)	Average Time (s)	Time Elapsed (s)
24	3.4	3.5	3.45	3.45
27	2.2	2.1	2.15	5.60
30	2.3	2.2	2.25	7.85
33	2.4	2.6	2.50	10.35
36	2.3	2.3	2.30	12.65
39	2.5	2.6	2.55	15.20
42	2.3	2.2	2.25	17.45
45	3.3	3.2	3.25	20.70
48	3.1	3.0	3.05	23.75
51	3.3	3.7	3.50	27.25
54	3.6	3.9	3.75	31.00
57	4.0	3.9	3.95	34.95
60	2.1	2.3	2.20	37.15
62	2.0	1.9	1.95	39.10
64	3.5	3.2	3.35	42.45
66	3.8	4.2	4.00	46.45
68	3.1	3.1	3.10	49.55
70	2.3	2.3	2.30	51.85
72	5.4	5.3	5.35	57.20
74	3.1	3.2	3.15	60.35
76	6.5	6.4	6.45	66.80
78	6.6	6.7	6.65	73.45
80	5.4	5.3	5.35	78.80
81	2.8	1.9	2.35	81.15
82	3.4	9.8	6.60	87.75
83	6.0	7.3	6.65	94.40
84	7.0	3.2	5.1	99.50
85	6.0	5.8	5.9	105.80

17.4 Data Reduction

The experimental data can be used to determine the heat transfer characteristics of the light bulb, including the fraction of the filament power absorbed by the glass bulb, the fractions of the heat transfer to the environment by convection and radiation and, by using correlations from the literature for natural convection and combined (i.e., natural and forced) convection, the heat transfer coefficients for natural and forced convection.

17.4.1 Glass Envelope Absorption Filament Radiation

To determine the absorptivity of the glass, a transient heat balance was written for the spherical bulb at time zero, shown in Equation 17.1:

$$Q_i - Q_o + Q_g = Q_a \tag{17.1}$$

Because $T_s = T_a$ at $t = 0$, the convection and radiation terms in Q_i and Q_o are zero. Thus, as is noted in Equation 17.2,

$$Q_g = Q_a \tag{17.2}$$

where Q_g is the heat absorbed by the glass as the thermal radiation from the filament passes through it and Q_a is the accumulation of heat in the glass. Thus, as is noted in Equation 17.3,

$$Q_g = mC_p \frac{dT_S}{dt} \tag{17.3}$$

where $\frac{dT_s}{dt}$ is the initial rate of bulb temperature rise (1.27 $\frac{K}{s}$ from Figure 17.4, in which the experimental data are plotted as T_s versus t), m is the bulb mass (13.24 g), and C_p is the glass heat capacity, 750 $\frac{J}{kg\,K}$. Then, $Q_a = Q_g = 12.6$ W.

The fraction of filament radiation absorbed by the glass envelope at the initial instant of heating is $f = \frac{Q_g}{Q_e} = \frac{12.6}{48} = 0.263$. At the initial instant of heating, the heat transfer by convection from the filament to the glass bulb is precisely zero.

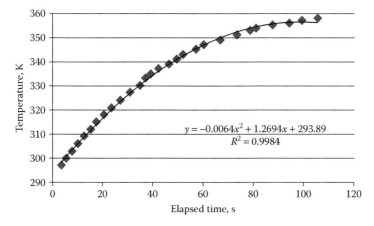

FIGURE 17.4
Plot of experimental data: Bulb surface temperature as a function of elapsed time.

17.4.2 Combined Convection Heat Transfer Coefficient

The determination of the combined convection heat transfer coefficient starts with a heat balance over the glass bulb at steady state, when Q_i and Q_a are both zero, as is shown in Equation 17.4:

$$Q_g = Q_o = Q_c + Q_r = h_c A_s (T_s - T_a) + \sigma \varepsilon A_s (T_s^4 - T_a^4) \tag{17.4}$$

The radiation heat transfer from the bulb to the classroom is 4.6 W, and $Q_g = Q_o = 12.6$ W, $Q_c = 8.0$ W, $D = 6$ cm, and $A_s = \pi D^2 = 0.011 \text{m}^2$. As a result, $h_c = 12.9 \frac{W}{m^2 K}$.

17.4.3 Natural Convection Heat Transfer Coefficient

The heat loss by combined convection is by natural and forced convection. The natural convection coefficient can be calculated by a literature correlation (Çengel and Ghajar 2011), shown in Equation 17.5:

$$\text{Nu}_n = \frac{h_n D}{k} = 2 + \frac{2 + 0.589 \text{Ra}^{0.25}}{\left[1 + \left(\frac{0.469}{P_r}\right)^{0.63}\right]^{0.444}} \tag{17.5}$$

In Equation 17.5, $k = 0.0274 \frac{W}{mK}$, $v = 1.85\text{E-}5 \frac{m^2}{s}$, $\beta = \frac{1}{327 K^{-1}}$, $Pr = 0.723$, $Gr = 1.4\text{E}6$, Ra = GrPr, and $h_n = 7.6 \frac{W}{m^2 K}$.

17.4.4 Forced Convection Heat Transfer Coefficient

In order to determine the forced convection heat transfer coefficient, a correlation for the combined coefficient as a function of h_n and h_f (Çengel and Ghajar 2011) must be used, as is noted in Equation 17.6:

$$\text{Nu}_c = (\text{Nu}_n{}^n + \text{Nu}_f{}^n)^{\frac{1}{n}} \rightarrow h_c{}^n = h_n{}^n + h_f{}^n \rightarrow h_f = (h_c{}^n - h_n{}^n)^{\frac{1}{n}} \tag{17.6}$$

$$(\textit{Note}: \textit{used } n = 3)$$

In Equation 17.6, $h_c = 12.9 \frac{W}{m^2 K}$, $h_n = 7.6 \frac{W}{m^2 K}$, and $h_f = 11.9 \frac{W}{m^2 K}$.
    Thus, the forced convection heat transfer coefficient was not measured. Is the resulting value reasonable for a classroom with an ambient temperature of 26°C (79°F)? To answer this question, it is essential to know what air velocity would yield this value of h_f. h_f can be determined from a forced convection correlation for spheres (Çengel and Ghajar 2011), shown in Equation 17.7:

$$\text{Nu}_f = 2 + [0.4\text{Re}^{\frac{1}{2}} 0.06\text{Re}^{\frac{2}{3}}]\text{Pr}^{0.4} \left(\frac{\mu_b}{\mu_s}\right)^{0.25} \tag{17.7}$$

with $\mathrm{Nu}_f = h_f D$, $\frac{h_f D}{k} = 26$, $\mathrm{Pr} = 0.723$, $\mu_b \approx \mu_s$, $\mathrm{Re} = \frac{VD}{V} = 2000$, and $V = 0.62 \frac{\mathrm{m}}{\mathrm{s}}$ (1.4 mph).

In the next section, the reasonableness of this value will be discussed.

17.5 Discussion of Results

Figure 17.4 is a plot of the bulb surface temperature, T_s, as a function of time, t. The data are fitted nicely with a second-order polynomial in T_s and t. As was noted earlier, the slope of the plot at time zero (i.e., $\frac{dT_s}{dt}$ in Equation 17.3) is 1.27 $\frac{K}{s}$.

Assuming free convection is negligible as a mode of heat transfer between the filament and the glass bulb, the fraction of heat radiated by the filament that is absorbed by the glass bulb is 0.263. This value could be verified independently by using Figure 12.9 from Çengel and Ghajar (2011), which presents the black body emissive power versus radiation wavelength for various source temperatures from 100 K to 5800 K, in conjunction with Figure 12.36 from Çengel and Ghajar (2011), which presents spectral transmissivity of low-iron glass as a function of radiation wavelength and glass thickness. These calculations were not carried out because all the other results of this experiment indicate that the measured fractional absorption of 26% is reasonable.

Table 17.2 summarizes the quantitative results of this experiment. The back-calculated forced convection coefficient is 160% of the natural convection coefficient. Is this reasonable? ASHRAE Standard 55P (2016) indicates that this is reasonable because Figure 5.2.3.1 of the proposed standard shows

TABLE 17.2

Summary of Experimental Results

Fractional absorptivity of the glass bulb	0.263
Heat transfer due to...	
Radiation, Q_r	4.6 W
Convection (natural + forced), Q_c	8.0 W
Percent of heat transfer due to...	
Radiation	36%
Convection	64%
Heat transfer coefficient for...	
Combined natural and forced convection, h_c	$12.9 \frac{W}{m^2 K}$
Natural convection, h_n	$7.6 \frac{W}{m^2 K}$
Forced convection, h_f	$11.9 \frac{W}{m^2 K}$
Velocity to produce h_f from correlation	$0.62 \frac{m}{s}$ (1.4 mph)

that an air speed of 0.6 $\frac{m}{s}$ (1.4 mph) will offset a room temperature increase of 7.2°C (4°F) above the operative comfort temperature of 23.9°C (75°F).

17.6 Conclusions

1. A simple but effective incandescent light bulb experiment, using readily available materials and equipment, was developed for use in the classroom and/or laboratory.

2. The fraction of filament radiation absorbed by the bulb was 26%. This value is reasonable based on the wavelengths of radiation produced by the element and based on other reasonable results of the experiment.

3. Forced convection in the classroom was greater than natural convection by a ratio of 1.6.

   From literature correlations, the air velocity required to produce the forced convection coefficient is 0.62 $\frac{m}{s}$ (1.4 mph). This value is within the ASHRAE guidelines for room air speeds allowed to offset room temperatures above the comfort zone temperature at a lower air speed of 0.4 $\frac{m}{s}$.

4. The experimental results indicate that, at steady state, the convection component of heat transfer from the filament to the bulb is very small relative to the radiation component.

17.7 Nomenclature

Latin Symbols

A_s	Surface area of the bulb, m²
C_p	Specific heat of glass, $\frac{W}{kg\,K}$
D	Bulb diameter, m
f	Fraction of the filament radiation that is absorbed by the glass envelope
g	Gravitational acceleration, $\frac{m}{s^2}$

h_c	Combined (natural + forced) convection heat transfer coefficient, $\frac{W}{m^2\,K}$
h_f	Forced convection heat transfer coefficient, $\frac{W}{m^2\,K}$
h_n	Natural convection heat transfer coefficient, $\frac{W}{m^2\,K}$
k	Thermal conductivity of air, $\frac{W}{mK}$
m	Mass of the bulb, kg
Q_a	Accumulation term in the heat balance equation, $= \frac{mC_p dT_s}{dt}$, W
Q_c	Heat loss from the glass bulb by combined natural and forced convection, W
Q_e	Measured power input to the light bulb from the electrical source, W
Q_g	Heat generation term in the heat balance equation, W
Q_i	Heat input term in the heat balance equation, W
Q_o	Heat output term in the heat balance equation, W
Q_r	Heat loss from the glass bulb by radiation to the classroom environment, W
t	Time measured from the start of power input to the light bulb, s
T_a	Ambient temperature, K; also, steady-state surface temperature, K
T_s	Bulb surface temperature, K
V	Air velocity required to produce h_f, $\frac{m}{s}$

Greek Symbols

β	Volume expansion coefficient, K^{-1}
μ	Kinematic viscosity, Pa s
ε	Surface emissivity of the tungsten filament
σ	Stefan–Boltzmann constant, $\frac{W}{m^2\,K^4}$
ν	Kinematic viscosity of air, $\frac{m^2}{s}$

Dimensionless Parameters

Gr	Grashof number, $\frac{g\beta(T_S - T_a)D^3}{\nu^2}$
Nu_c	Nusselt number, $\frac{h_e D}{k}$, combined natural and forced convection
Nu_f	Nusselt number, forced convection
Nu_n	Nusselt number, natural convection
Re	Reynolds number, $\frac{VD\rho}{\mu}$, required to produce the forced convection coefficient, h_f
Ra	Rayleigh number, GrPr

References

ASHRAE. 2016. ASHRAE Standard 55 P, *Thermal Environmental Conditions for Human Occupancy*. Accessed August 18, 2017. Available through https://www.ashrae.org/

Çengel, Y.A. and Ghajar, A.J. 2011. *Heat and Mass Transfer, Fundamentals and Applications*, 4th ed. New York: McGraw-Hill.

Hammack, B. 2012. *Light Bulb Filament*. Mr. Barlow's Blog. Accessed August 18, 2017. http://mrbarlow.wordpress.com/2011/03/16/interesting-facts-about-light-bulb-filament.

Riveros, H.G. and Oliva, A.I. 2006. Experiment tests pressure in light bulbs. *Physics Education*, 487–486, 2006. Available at http://www.mda.cinvestav.mx/labs/fisica/micros/laboratorio/articulos%20pdf/Internacionales/67.pdf.

18

Experimental Determination of a Room Heat Transfer Coefficient by Transient Cooling of a Mercury-in-Glass Thermometer

William Roy Penney and Edgar C. Clausen

CONTENTS

18.1 Introduction

Transient convection is very important in a variety of industrial and environmental settings including human comfort in buildings, atmospheric flow, thermal regulation processes, the cooling of electronic devices, and maintaining the security of energy systems (Padet 2005). Even our everyday comfort depends on transient heat transfer analysis to ensure that a comfortable temperature range is maintained in buildings.

Unavoidable forced convection in all occupied spaces—including laboratory rooms—provides convection coefficients that are significantly greater than those obtained strictly by free convection. Thus, accurate modeling must include a combination of free and forced convection, commonly identified as mixed or combined convection (Çengel and Afshin 2011a).

The purpose of this experiment was to:

1. Experimentally determine the transient temperature versus time of a cooling horizontal mercury-in-glass thermometer.
2. Develop a mathematical model to predict the transient temperature versus time, considering only natural convection and radiation.
3. Develop a mathematical model to predict the transient temperature versus time, considering combined forced and natural convection and radiation.
4. Determine the theoretical forced convection coefficient by minimizing the standard deviation between the mathematical model predictions to the experimental data.

18.2 Experimental

18.2.1 Equipment List

The following equipment was required to carry out the experiment:

- Thermometer, mercury-in-glass, 10 mm-diameter bulb
- Blow dryer, Sunbeam, Model 1875
- Ring Stand, fitted with clamp
- Clamp
- Digital Caliper, Cen-Tech 47257
- Stop watch, VWR

18.2.2 Experimental Apparatus

Figure 18.1 shows a photograph of the experimental apparatus, including a digital caliper, a ring stand fitted with a clamp, a mercury-in-glass thermometer and a blow dryer.

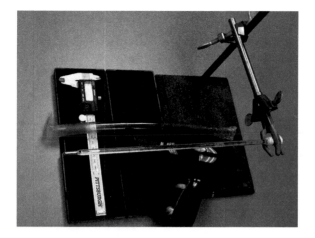

FIGURE 18.1
Photograph of the experimental apparatus.

18.2.3 Experimental Procedure

18.2.3.1 Experimental Setup

1. Setup the experimental apparatus as shown in Figure 18.1.
2. Turn on the digital caliper, change the display to mm, and zero the caliper.
3. Measure the length and diameter of the mercury-in-glass thermometer bulb using the digital caliper.
4. Clamp the end of the thermometer to hold the bulb in a horizontal position in mid-air.
5. Measure the ambient room temperature.

18.2.3.2 Experimental Procedure

Recruit three students to perform the experiment:

1. The first student uses the blow dryer to heat the bulb of the thermometer to an initial temperature slightly over 60°C.
2. After the first student moves the blow dryer to a safe location, the second student waits until the temperature reads exactly 60°C, and starts the stopwatch. The second student then calls out the times at temperature readings of 50°C, 40°C, 35°C, 32°C, 30°C, 28°C, 26°C, 25°C, and 24°C.
3. The third student records the experimental temperature and time data as it is called out by the second student.

18.2.3.3 Safety

Care should be taken to not break the mercury-in-glass thermometer. Acute exposure to mercury is toxic.

18.3 Raw Experimental Data

The experimental data of temperature as a function of time is shown in Table 18.1. The room temperature was measured as 22°C.

18.4 Reduction of Experimental Data

18.4.1 Developing the Transient Model

Begin with a heat balance over the system of the thermometer bulb (Çengel and Afshin 2011b), shown in Equation 18.1:

$$Q_i - Q_o + Q_g = Q_a \qquad (18.1)$$

Since there is no heat entering the system, $(Q_i = 0)$ or generated $(Q_g = 0)$ within the system, the heat balance reduces to Equation 18.2:

$$-Q_o = Q_a \qquad (18.2)$$

TABLE 18.1

Experimental Data

Time, s	Temperature, °C
0	60
21.02	50
74.02	40
113.55	35
144.92	32
175.08	30
209.80	28
259.20	26
303.73	25
359.98	24

The heat accumulated within the bulb system is calculated by Equation 18.3:

$$Q_a = mC_p \frac{dT_s}{dt} \tag{18.3}$$

For the case of natural convection only, the heat out of the system is by natural convection and radiation (Çengel and Afshin 2011b), shown in Equation 18.4:

$$Q_o = h_n A_s (T_s - T_a) + \sigma \varepsilon A_s (T_s^4 - T_a^4) \tag{18.4}$$

For the case of natural convection and forced convection, the heat out of the system is by natural convection, forced convection, and radiation (Çengel and Afshin 2011b), shown in Equation 18.5:

$$Q_o = h_c A_s (T_s - T_a) + \sigma \varepsilon A_s (T_s^4 - T_a^4) \tag{18.5}$$

Substituting Equations 18.3 and 18.4 into Equation 18.2 yields Equation 18.6:

$$-h_n A_s (T_s - T_a) - \sigma \varepsilon A_s (T_s^4 - T_a^4) = mC_p \frac{dT_s}{dt} \tag{18.6}$$

Substituting Equations 18.3 and 18.5 into Equation 18.2 yields Equation 18.7:

$$-h_c A_s (T_s - T_a) - \sigma \varepsilon A_s (T_s^4 - T_a^4) = mC_p \frac{dT_s}{dt} \tag{18.7}$$

Solving Equation 18.6 for the change in the bulb surface temperature with respect to time yields Equation 18.8:

$$\frac{dT_s}{dt} = \frac{\left[-h_n A_s (T_s - T_a) - \sigma \varepsilon A_s (T_s^4 - T_a^4) \right]}{mC_p} \tag{18.8}$$

Solving Equation 18.7 for the change in bulb surface temperature with respect to time yields Equation 18.9:

$$\frac{dT_s}{dt} = \frac{\left[-h_c A_s (T_s - T_a) - \sigma \varepsilon A_s (T_s^4 - T_a^4) \right]}{mC_p} \tag{18.9}$$

Equations 18.8 and 18.9 can be used to calculate the change in the surface temperature of the bulb with respect to time for the cases of natural convection only and combined convection, respectively. Before Equations 18.8 and 18.9 can be utilized, a number of other parameters must be calculated.

The surface area of the thermometer bulb is calculated by Equation 18.10:

$$A_s = \pi D L \tag{18.10}$$

Neglecting the glass bulb container, the mass of the mercury in the thermometer bulb is calculated by Equation 18.11:

$$m = \rho_{hg}\left(\frac{\pi}{4}\right)D^2L \tag{18.11}$$

Heat transfer from the surface of the thermometer bulb is partially by natural convection. To determine the value of the natural convection heat transfer coefficient, the natural convection Nusselt number must first be calculated (Çengel and Afshin 2016c), shown in Equation 18.12:

$$Nu_n = \frac{\left(0.6 + 0.387 Ra^{\frac{1}{6}}\right)}{\left(1 + \left(\frac{0.492}{Pr}\right)^{\frac{9}{16}}\right)^{\frac{8}{27}}} \tag{18.12}$$

This expression for the natural convection Nusselt number involves a number of unknowns that have to be calculated first: the Rayleigh number and the Prandtl number for the air around the thermometer bulb. The Rayleigh number is shown in Equation 18.13 (Çengel and Afshin 2011d):

$$Ra = \frac{g\beta(T_s - T_a)D^3}{\upsilon^2} \tag{18.13}$$

The Prandtl number of the air around the bulb is defined in Equation 18.14 as (Çengel and Afshin 2011b):

$$Pr = \mu\frac{C_p}{\rho} \tag{18.14}$$

These equations rely on a number of other parameters, $\beta, \upsilon, \mu,$ and ρ (Çengel and Afshin 2011e), shown in Equation 18.15:

$$\beta = \frac{1}{T_f} \tag{18.15}$$

A curve fit of v versus T_f yields Equations 18.16 and 18.17:

$$\upsilon = 1.34 \times 10^{-5} + (8.81 \times 10^{-8})T_f - (9.00 \times 10^{-11})T_f^2 \tag{18.16}$$

$$\rho = \frac{Pm_w}{(RT_f)} \tag{18.17}$$

T_f is the film temperature of the air surrounding the thermometer bulb at each time (Çengel and Afshin 2011d), as noted in Equation 18.18:

$$T_f = \frac{(T_s + T_a)}{2} \tag{18.18}$$

Now, Nu_n can be calculated. The natural convection heat transfer coefficient is then calculated by (Çengel and Afshin 2011d), shown in Equation 18.19:

$$h_n = \frac{Nu_n k}{D} \tag{18.19}$$

Given a forced convection heat transfer coefficient (h_f), a forced Nusselt number can be calculated, as in Equation 18.20:

$$Nu_f = \frac{h_f D}{k_a} \tag{18.20}$$

A combined Nusselt number can be calculated from the natural and forced convection Nusselt numbers. The combined Nusselt number can be used to calculate the combined heat transfer coefficient (Çengel and Afshin 2011f), as noted in Equations 18.21 and 18.22:

$$Nu_c = (Nu_f^n + Nu_n^n)^{1/n} \tag{18.21}$$

$$h_c = \frac{(Nu_c k)}{D} \tag{18.22}$$

Now the temperature of the surface of the bulb with respect to time can be calculated for the case of natural convection and radiation by Equation 18.8 and for the case of natural convection, forced convection, and radiation by Equation 18.9 using a simple Euler integration technique.

18.4.2 Statistical Analysis

The relative standard deviation between the data sets for experimental temperature and the theoretical temperature versus time was determined (Zamboni 2017), as shown in Equation 18.23:

$$\sigma_{SD} = \sqrt{\left(\frac{1}{N}\right)\sum_{i=1}^{N}\left(\frac{(T_m - T_e)}{(T_m)}\right)^2} \tag{18.23}$$

In order to calculate the relative standard deviation between the experimental temperature and the theoretical temperature versus time data, the temperatures at each experimental time for the theoretical temperature versus time data must be calculated. This was accomplished by curve fitting the entire theoretical temperature versus time model to determine regression coefficients. With the regression coefficients, the theoretical temperatures (T_m) corresponding to the experimental temperatures (T_e) at each experimental node can be calculated by Equation 18.24:

$$T_m = b_0 + b_1(t_e) + b_2(t_e)^2 \tag{18.24}$$

where b_0, b_1, and b_2 are the polynomial regression coefficients for the curve fit of the model-predicted temperature versus time data. With T_m, the standard deviation can now be calculated.

The percent difference at each time between the experimental temperature versus time data and the theoretical temperature versus time are shown in Equation 18.25 (Math Is Fun 2014):

$$\%_D = \frac{(T_e - T_m)}{\left(\frac{(T_e + T_m)}{2}\right)} \times 100 \tag{18.25}$$

The average percent difference between the experimental temperature distribution and the theoretical temperature distribution for all nodes along the fin are shown in Equation 18.26 (Math Is Fun 2014):

$$\%_{D_{avg}} = \left(\left(\frac{1}{N}\right) \sum_{i=1}^{N} \%_D \right) \times 100 \tag{18.26}$$

18.5 Reduced Results

The results of the model that best fit the experimental transient temperature data are summarized in Table 18.2. Figure 18.2 is a plot of the experimental transient temperature data and the transient temperature data predicted by the mathematical model with and without forced convection for a mercury-in-glass thermometer.

18.6 Discussion of Results

The results of this experiment indicate that there was forced convection in the room where the mercury-in-glass thermometer cooling experiment was conducted. The mathematical model developed without forced convection

TABLE 18.2

Model-Predicted Results from the Mathematical Model

Time, s	Temperature, °C
0	59.20
26.02	51.73
74.02	41.12
113.55	35.01
144.92	31.54
175.08	29.14
209.80	27.25
259.20	25.61
303.73	24.58
359.98	22.87

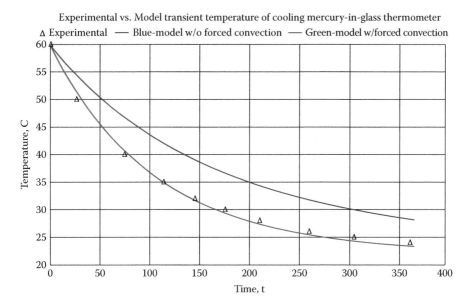

FIGURE 18.2

Plot of the transient temperature versus time for the experimental data and for the mathematical model with and without forced convection.

predicted very slow cooling relative to the experimental data. The mathematical model developed with forced convection, using a forced convection coefficient of $18 \frac{W}{m^2K}$, the value that best fit the experimental data, gave an excellent agreement between the model predictions and the experimental data.

This *best fit* forced convection coefficient of $18 \frac{W}{m^2K}$ minimized the relative standard deviation to 2.62%. The average percent difference between the experimental temperature and the model-predicted temperature was 2.27%.

18.7 Conclusions

1. The relative standard deviation of the experimental data and the mathematical model was minimized at 2.62% for a forced convection coefficient of $18 \frac{W}{m^2K}$. This indicates there are significant forced convection effects present in the room.

2. The percent difference between the experimental data and the mathematical model was calculated to be 2.27%, which indicated that the mathematical model fit the experimental data very closely.

3. The forced convection present in the room is likely due to the ventilation and air conditioning system within the classroom.

18.8 Recommendations

1. Conduct additional runs, paying special attention to proper procedure to mitigate human error. Duplicate all of the runs.

2. Minimize movement within the classroom while the experiment is being conducted.

3. Use an enclosure within the laboratory room that eliminates all forced convection so that the model using only free convection can be adequately tested with experimental data.

4. Calibrate a thermocouple to accurately measure the same temperature as that measured by the thermometer. Extrapolate the combined heat transfer coefficient to $\Delta T = 0$, where the free convection coefficient is zero and all the heat transfer is by forced convection.

18.9 Nomenclature

Latin Letters

A_s	Surface area of thermometer bulb, m^2
C_p	Heat capacity, $\frac{J}{kg\,K}$
D	Bulb diameter, m
$\frac{dT_s}{dt\,T_m}$	Change in bulb and its surface temperature with respect to time, $\frac{K}{s}$
h_c	Combined convection heat transfer coefficient, $\frac{W}{m^2 K}$
h_n	Natural convection heat transfer coefficient, $\frac{W}{m^2 K}$
h_f	Forced convection heat transfer coefficient, $\frac{W}{m^2 K}$
k	Thermal conductivity of air, $\frac{W}{m\,K}$
L	Length of bulb, m
m_w	Molecular weight of air, $\frac{kg}{kg\text{-mol}}$
Nu_n	Natural convection Nusselt number, dimensionless
Nu_f	Forced convection Nusselt number, dimensionless
Nu_c	Combined convection Nusselt number, dimensionless
P	Pressure, Pa
Pr	Prandtl number, dimensionless
Q_a	Heat accumulation, W
Q_g	Heat generation, W
Q_i	Heat transfer into the bulb, W
Q_o	Heat transfer out of the bulb, W
R	Ideal gas constant, $\frac{J}{kg\text{-mol}\,K}$
Ra	Rayleigh number, dimensionless
T	Time, s
T_a	Ambient temperature, K
T_f	Film temperature, K
T_s	Surface temperature of the bulb, K
T_t	Theoretical temperature at time t, s
M	Mass of mercury in the thermometer bulb, kg
N	Number of experimental temperature versus time data points, dimensionless
T_m	Curve-fitted theoretical temperatures evaluated at the time of each experimental data point, C
T_e	Experimental temperature at each time, C
t_e	Experimental times, s

$\%_D$	Percent difference between the experimental data and the data predicted by the model, dimensionless
$\%_{D_{avg}}$	Average percent difference between the experimental data set and the set of data predicted by the model, dimensionless.

Greek Letters

β	Volume expansion coefficient, $\frac{1}{K}$
ε	Emissivity of mercury (Omega 2017), dimensionless
ν	Kinematic viscosity of air, $\frac{m^2}{s}$
σ	Stefan–Boltzmann constant, $\frac{W}{m^2 K^4}$
μ	Dynamic viscosity of air, $\frac{kg}{ms}$
ρ	Density of air, $\frac{kg}{m^3}$
ρ_{Hg}	Density of mercury, $\frac{kg}{m^3}$
σ_{SD}	Standard deviation between experimental and theoretical temperature versus time data, dimensionless

---

Acknowledgment

This chapter contains results from laboratory reports prepared by former University of Arkansas, Ralph E. Martin Department of Chemical Engineering, students including Brett A. Bland and Alexander R. Enderlin. The authors are extremely grateful for the hard work and dedication of these University of Arkansas graduates.

References

Çengel, Y.A. and Afshin, G.J. 2011a. *Heat and Mass Transfer*, 4th ed., p. 562. New York: McGraw-Hill.

Çengel, Y.A. and Afshin, G.J. 2011b. *Heat and Mass Transfer*, 4th ed. New York: McGraw-Hill.

Çengel, Y.A. and Afshin, G.J. 2011c. *Heat and Mass Transfer*, 4th ed., p. 542. New York: McGraw-Hill.

Çengel, Y.A. and Afshin, G.J. 2011d. *Heat and Mass Transfer*, 4th ed., p. 541. New York: McGraw-Hill.

Çengel, Y.A. and Afshin, G.J. 2011e. *Heat and Mass Transfer*, 4th ed., p. 536. New York: McGraw-Hill.

Çengel Y.A. and Afshin, G.J. 2011f. *Heat and Mass Transfer,* 4th ed., p. 563. New York: McGraw-Hill.

Math Is Fun. 2014. *Percentage Difference.* Accessed August 21, 2017. http://www.mathsisfun.com/percentage-difference.html.

Omega. 2017. *Emissivity of Common Materials.* Accessed August 21, 2017. http://www.omega.co.uk/literature/transactions/volume1/emissivitya.html.

Padet, J. 2005. Transient convective heat transfer. *Journal of the Brazilian Society of Mechanical Science and Engineering* 27(1) 74–95. Accessed August 21, 2017. Available at http://www.scielo.br/pdf/jbsmse/v27n1/25376.pdf.

Zamboni, J. 2017. *How to Calculate Percent Deviation.* Sciencing June 1. Accessed August 21, 2017. http://www.ehow.com/how_6384591_calculate-relative-deviation.html.

19

Experimental Determination and Modeling of the Temperature Distribution Along a Hollow Fin

William Roy Penney and Edgar C. Clausen

CONTENTS

19.1 Introduction

The use of fins to improve the performance of heat exchangers is very important; a search of *heat transfer fins* yields about 4,190 results on YouTube and about 654,000 results on Google. Mathematical modeling of fin performance is a very useful design tool because fin performance can be determined from fundamental heat transfer principles; thus, an experimental approach is not needed. Experiments aimed at determining the temperature distribution within fins and a comparison of the simulation results with experimental results are a very beneficial learning experience to help the student understand the fundamental principles involved in the understanding of fins and the modeling needed to predict fin performance.

The purpose of this study was to:

1. Experimentally determine the temperature profile along a hollow fin subjected to forced convection of room air.
2. Measure the air velocity across the fin and use a literature correlation to determine the forced convection heat transfer coefficient.
3. Develop a mathematical model to calculate the temperature profile along the fin.
4. Compare the model-determined results with experimental data.

19.2 Experimental

19.2.1 Equipment Description

The following equipment was used in the execution of this experiment:

- Hot water reservoir (i.e., a stainless steel cylinder with a welded-on bottom plate), 10.2 cm i.d., 11.4 cm o.d. 6.4 mm thickness, 21.6 cm height (4 in i.d., $4\frac{1}{2}$ in o.d., $\frac{1}{4}$ in thickness, $8\frac{1}{2}$ in height)
- Hollow copper fin, 4.0 mm i.d., 6.4 mm o.d., 17.8 cm length ($\frac{3}{16}$ in i.d., $\frac{1}{4}$ in o.d., 7 in length)
- Thermocouple reader, Omega Model HH12A
- Thermocouple, Omega, Type K, 1.6 mm o.d., sheathed, 29.8 cm long ($\frac{1}{16}$ in o.d., $11\frac{3}{4}$ in long)
- Air flow meter, Kane–May, Model #KM4107
- Automerse heater, Fisher, Model #199
- Bubble wrap insulation, 21.6 × 50.8 cm ($8\frac{1}{2}$ × 20 in)
- Fan, Lasko, Three-Speed Breeze Machine
- Yard Stick

19.2.2 Experimental Apparatus

Photographs of the experimental apparatus are shown in Figures 19.1 through 19.3. The apparatus was assembled with all the supplies listed in the previous equipment description list.

FIGURE 19.1
Photograph of the experimental apparatus.

FIGURE 19.2
Photograph of the Automerse heater control.

FIGURE 19.3
Zoomed-out photograph of the experimental apparatus.

19.2.3 Experimental Procedure

19.2.3.1 Experimental Setup

1. Prior to setup, the hollow copper fin was welded to the lower portion of the hot water reservoir.
2. Wrap insulation around the hot water reservoir.
3. Fill the hot water reservoir with water.
4. Place the Automerse heater in the cylinder.
5. Connect the thermocouple leads to the thermocouple reader and turn it on. Set it to read the temperature in degrees Celsius.
6. Place the fan approximately 125 cm (50 in) away from the cylinder and fin. Turn the fan on to the lowest setting and face it towards the cylinder and fin.
7. Recruit two people to conduct the experiment.
   a. One person to monitor the heater and to take temperature readings.
   b. One person to record the data from the experiment.

19.2.3.2 Experimental Procedure

1. Turn the heater setting to 10.
2. Allow the water to heat to 100°C. Turn the heater setting to 8 to maintain the boiling of the water.

3. Insert the thermocouple into the hollow fin until the thermocouple bottoms out against the cylinder wall.

4. Record the temperature and distance from the cylinder.

5. Move the thermocouple tip 2.5 cm (1 in) farther out of the hollow fin.

6. Record the temperature and distance from the cylinder.

7. Repeat steps 6 and 7 until the end of the fin is reached.

8. Use the air flow meter to measure the air velocity at the base of the fin and in four evenly spaced increments along the fin.

19.2.3.3 Safety

1. Safety glasses, closed-toed shoes, and long pants must be worn during the experiment.

2. Use caution to not touch hot surfaces when operating the Automerse heater.

3. The stainless steel cylinder and copper fin are hot. Be aware of the equipment temperature and avoid touching hot surfaces.

19.3 Raw Experimental Data

The raw experimental data are shown in Table 19.1.

TABLE 19.1

Raw Experimental Data

Length from Base, in	Temperature, °C	Air Velocity, $\frac{ft}{min}$
0	80.0	400
0.5	69.3	—
1.0	61.0	—
1.75	—	370
2.0	50.5	—
3.0	43.5	—
3.5	—	360
4.0	38.5	—
5.0	35.0	—
5.25	—	410
5.5	32.5	—
6.0	31.0	—
7.0	—	410

19.4 Reduction of Experimental Data: Determining the Heat Transfer Coefficient for Crossflow over a Cylinder

The Churchill and Bernstein equation (Çengel and Ghajar 2015a) was used, as is noted in Equation 19.1:

$$\text{Nu} = \frac{h_m D}{k_a} = 0.3 + \frac{0.62 \text{Re}^{\frac{1}{2}} Pr^{\frac{1}{3}}}{\left[1 + \left(\frac{0.4}{Pr}\right)^{\frac{2}{3}}\right]^{\frac{1}{4}}} \left[1 + \left(\frac{\text{Re}}{282,000}\right)^{\frac{5}{8}}\right]^{\frac{4}{5}} \tag{19.1}$$

The Reynolds number was calculated from Çengel and Ghajar (2015b) and is shown in Equation 19.2:

$$Re = \frac{VD\rho}{\mu} \tag{19.2}$$

The heat transfer coefficient for radiation is given by Çengel and Ghajar (2015c) and is shown in Equation 19.3:

$$h_r = \frac{\varepsilon\sigma\left(T_x^4 - T_a^4\right)}{T_x - T_a} \tag{19.3}$$

The radiation heat transfer coefficient was evaluated midway along the fin.

Once the heat transfer coefficients are known for both radiation and convection, the total heat transfer coefficient can be calculated, as is shown in Equation 19.4:

$$h_{\text{tot}} = h + h_{ri} \tag{19.4}$$

The parameters m, p, and A_c are needed to determine the temperature distribution along the fin. Equations for these parameters are shown in Equations 19.5 through 19.7:

$$p = \pi D \tag{19.5}$$

$$A_c = \frac{\pi}{4(D^2 - D_i^2)} \tag{19.6}$$

$$m = \frac{h_{\text{tot}} p}{k A_c} \tag{19.7}$$

The fin is sufficiently long so that the heat transfer from the tip can be ignored; thus, the temperature distribution along the fin is given by Çengel and Ghajar (2015d) and is shown in Equation 19.8:

$$\frac{T_x - T_a}{T_b - T_a} = \frac{\cos[m(L-x)]}{\cosh[mL]} \tag{19.8}$$

19.5 Reduced Results

Figure 19.4 presents the model results compared with experimental data. The average deviation of the model-predicted temperatures from the experimental temperatures is about 2.5%.

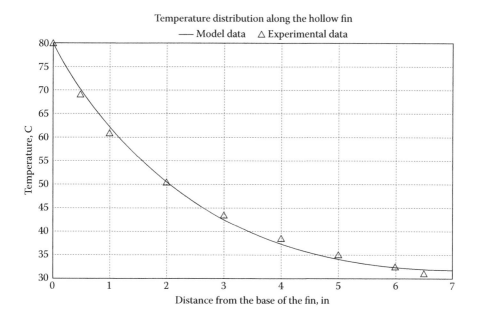

FIGURE 19.4
A comparison of the fin temperature profile from the mathematical model with the experimentally determined profile.

19.6 Discussion of Results

The mathematical model predicts the experimental data very well. The experimental temperature at the tip of the fin is too low because (1) the thermocouple tip is too close to the room air and (2) there is a relatively large conduction error for the thermocouple because the tip is very close to the location along the thermocouple, where contact with room air occurs.

19.7 Conclusions

1. A hollow fin is an excellent choice for measuring the fin temperature by inserting a thermocouple into the hollow of the fin.

2. The temperature profile predicted by a mathematical model, taken from the literature, fits the experimental data with an absolute deviation of about 2.5%.

3. The experimental procedures are very adequate to obtain accurate experimental temperature profiles.

19.8 Recommendations

1. Duplicate experiments could be used to assess the experimental accuracy.

2. A small thermocouple attached to the fin tip would improve the accuracy of the tip temperature measurement.

3. A copper or aluminum hot water reservoir would give base temperatures closer to the 100°C temperature of the boiling water.

4. A transient experiment could be conducted and modeled by

   - Pouring boiling water into the reservoir at the start of the experiment.
   - Recording transient temperatures at several locations along the fin with time.
   - Curve fitting the base temperature transient profile and including it in a transient nodal mathematical model.

19.9 Nomenclature

Latin Symbols

A_c	Cross-sectional area of the fin, m²
A_s	Surface area of the fin, in²
D	Outer diameter of the fin, m
D_i	Inner diameter of the fin, m
Re	Reynolds number, unitless
h_c	Convection heat transfer coefficient, $\frac{W}{m^2 K}$
h_r	Heat transfer coefficient due to radiation, $\frac{W}{m^2 K}$
h_{tot}	Total heat transfer coefficient, $\frac{W}{m K}$
k_a	Air thermal conductivity, $\frac{W}{m K}$
k_f	Fin thermal conductivity, $\frac{W}{m K}$
L	Fin length, in
m	$\frac{h_{tot} p}{(k A_c)}$, unitless
Nu	Nusselt number, unitless
p	πD, Perimeter of the fin, m
V_{avg}	Average velocity of air, $\frac{m}{s}$
T_{bn}	Temperature at the base of the fin, C
T_a	Temperature of ambient the air, C
T_x	Temperature of the fin at distance x from its base, C
V	Velocity, $\frac{m}{s}$
x	Length along the fin from its base, m

Greek Symbols

ρ	Density, $\frac{kg}{m^3}$
μ	Viscosity, $\frac{kg}{m s}$

Acknowledgment

This chapter contains results from laboratory reports prepared by former University of Arkansas, Ralph E. Martin Department of Chemical Engineering, students including Kali M. McGhehey and Andrea M. Dieterle. The authors are extremely grateful for the hard work and dedication of these University of Arkansas graduates.

References

Çengel, Y.A. and Ghajar, A.J. 2015a. *Heat and Mass Transfer—Fundamentals and Applications,* p. 442, 5th ed. New York: McGraw-Hill.

Çengel, Y.A. and Ghajar, A.J. 2015b. *Heat and Mass Transfer—Fundamentals and Applications,* p. 441, 5th ed. New York: McGraw-Hill.

Çengel, Y.A. and Ghajar, A.J. 2015c. *Heat and Mass Transfer—Fundamentals and Applications,* p. 145, 5th ed. New York: McGraw-Hill.

Çengel, Y.A. and Ghajar, A.J. 2015d. *Heat and Mass Transfer—Fundamentals and Applications,* p. 174, 5th ed. New York: McGraw-Hill.

Index

Note: Page numbers followed by f and t refer to figures and tables respectively.